Social Blend

Helping Restauranteurs leverage WordPress
to create an online presence
while engaging their community and growing their
RESTAURANT *&* **CATERING** *business*

Marna Friedman

Cover artwork: © Valdgrin/Shutterstock, ©depositphotos/artjazz

SOCIAL Blend
Published by MWF Publishing, Dallas, GA

OTHER BOOKS BY MARNA FRIEDMAN
EVENtually Perfect
The Small Business That Could™
The Small Business That Could™ For Women
The Social Launch Toolkit
Real Smarts: Leveraging smartphones and new technologies in your Real Estate Business
Agent Smarts: Real Estate Websites Made With WordPress

ISBN: 978-0-9840169-4-5
Library of Congress Control Number: 2013900432

Manufactured in the United States of America

Table of Contents

Symbol Key:

There are several items that I wanted to call out in this book. They are each broken out into specific types of categories:

Technology tidbit

Definitions

Restaurant Detail

Why WordPress

⚜ CHAPTER ⚜

1

Learn how to create a WordPress website
📖 WordPress as a CMS
📖 Advantages of WordPress

WordPress as a CMS

While there are many options available for your website platform, a content management system (CMS) allows you to easily update your content. WordPress also enables you to quickly launch a website for your restaurant and/or catering business and plugins help you expand the functionality of your website.

NOTES

_____

_____

_____

_____

Technology

A CMS like WordPress enables you to own and control your website, content and links.

Unfortunately, most restaurant & catering professionals do not have a website, and this can be detrimental to their business.

It's no secret that the restaurant market has moved online and mobile. Last year, over 90% of consumers started their search for a restaurant on the internet, while over 50% used the same medium to also research restaurant reviews. With these kind of statistics, it's vital to have an effective, professional restaurant website to compete.

Most chain restaurants provide websites to their locations and/ or franchisees. This can be a quick and easy way for a restaurant to establish themselves online. But smaller, independent restaurants struggle to be found by new customers.

Another issue for restaurants is the learning curve. Most "packaged" restaurant websites, whether from a corporate location or purchased through another company typically promise training and support, which to a restaurant owner struggling to find ex-

NOTES

_____

_____

_____

_____

tra time, can be quite appealing. But this too, comes with issues. The restauranteur needs to learn the platform and when they can't find time to accomplish this, the site usually just becomes an outdated portal of basic menu items, some of which may no longer be available, and static information.

Ask your typical restauranteur about WordPress, and most won't know what you're talking about. If they have heard of Word-Press, chances are they think it's simply a blogging platform, maybe something to use in conjunction with their main site, if at all. However, WordPress can offer'; the kind of unmatched flexibility, affordability and power to provide a comprehensive, all-in-one restaurant solution for even the most demanding of restauranteurs.

It's important to understand how a content management system can help with your restaurant and catering business strategy. And using WordPress as your CMS offers tremendous possibilities. According to Forbes magazine, in 2012 WordPress powered one of every 6 websites on the Internet, nearly 60 million in all, with 100,000 more popping up each day. And today, according to w3techs, WordPress is used by 59.1% of all the websites whose content management system we know. This is 20.4% of all websites.

Advantages of WordPress

There are several reasons why WordPress continues to be the CMS of choice, but the major reasons are:

Notes

_____

_____

_____

_____

SEO

Everyone wants their website to be found by search engines, and they frequently struggle with search engine optimization, sometimes spending a significant amount of money on it. But WordPress offers clean code that is properly formatted and organized in a simple way, as well as offering SEO friendly URLs that enable every page to have a distinctive title and meta description.

Easy Set-up and Launch

WordPress was developed to be easy, and once installed can be used immediately. For those that are hesitant, they can practice with a WordPress.com website and learn the basics before launching a self-hosted site. And then if they want something more unique, it is quite easy to find a developer to help with that. A system as popular as WordPress offers access to help, whether it be local groups, hiring a developer, or seeking more formal education.

Themes

Another plus related to the widespread affinity for WordPress is that custom, unique templates are easy to find and sometimes free. There are also an endless number of independent designers out there who create SEO friendly templates.

Options to Optimize

Adding plugins to your site is helpful if you're looking to increase its functionality. Before making any changes, ensure that they're SEO friendly and will help instead of hurt when it comes to local search.

Notes

_____

_____

_____

_____

In the Know

Many people will tell you how easy WordPress is, and compared to many other web platforms, this is true. But setting up a website may not be your thing. Be mindful of what you do best and remember why people choose you as their restaurant. It's OK to be good at some things and not others. Knowing your strengths is great. Knowing when to seek help is even better. Explore what's possible and then make an educated decision.

NOTES

_____

_____

_____

_____

NOTES

_____

_____

_____

_____

WordPress.com for Restaurants

Learn about the Restaurant website options available on WordPress.com

- 📖 Creating a WordPress.com website
- 📖 Restaurant Functionality
- 📖 Upgrade Options

WordPress.com is typically not what I recommend for a business website since it does not offer the ability to add plugins, edit CSS, or several other features. But last year they created a special program for restaurant websites offering a "package" of options including:

- Hours & Location
- Simple Menus
- Locu Menu Embed
- Open Table Reservation integration
- Responsive websites that are mobile ready

NOTES

_____

_____

_____

_____

- restaurant themes

In addition, you can upgrade your website with some premium features including:
- Custom domain
- Storage
- Video
- Custom design

The sites are easy to implement and set-up.

Confit & Bon Vivant: New Themes for Restaurants

You want a stylish restaurant site that sets up quickly, is simple to maintain, and gives your customer mobile access. It's a tall order — and our restaurant themes are built to satisfy. Easily add your location and hours, menu items, and a background photo for a tasteful site you can call your own.

This is probably a great way for a new restaurant to get started, although I would recommend that you use your own domain, and not http://joesrestaurant.com/wordpress. And as I stated earlier, one of the benefits of this program is your ability to add a custom domain.

NOTES

_____

_____

_____

_____

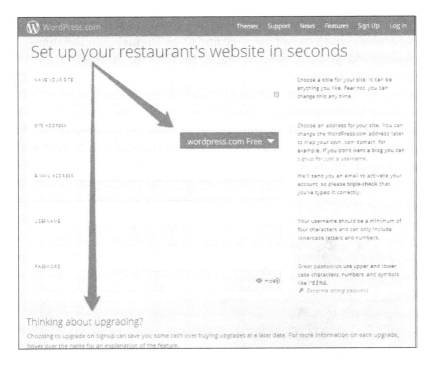

Note the dropdown at **WordPress.com FREE** - you can see what other options would cost you and make a selection. They also remind you about upgrading and offer you an opportunity to review the options available to you through this program. Once you make all of your selections and register your site, you can begin to navigate the features and build your website.

NOTES

When you are ready to select a theme, you will need to navigate on your dashboard to Appearance and then Themes. On the themes dashboard, you will need to search for Restaurant themes. There are currently two themes being offered through this program, one of which is a premium theme. Since you do not have the ability to add plugins to your website, you will need to select one of these in order to take advantage of the restaurant functionality being offered.

Once you make a selection, your WordPress dashboard will change to include the custom post type features of your website, i.e., food menus and widgets for Open Table, Hours, etc. You can navigate the dashboard to familiarize yourself with these features before you begin to build your website, as you can see on the following page.

NOTES

_____

_____

_____

_____

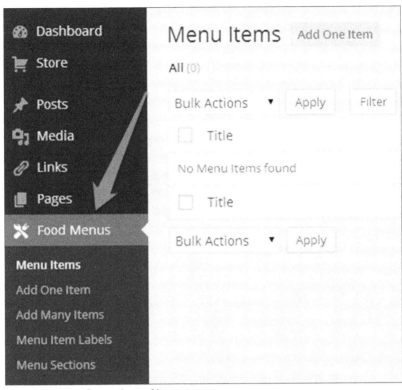

food menu functionality

NOTES

_____

_____

_____

_____

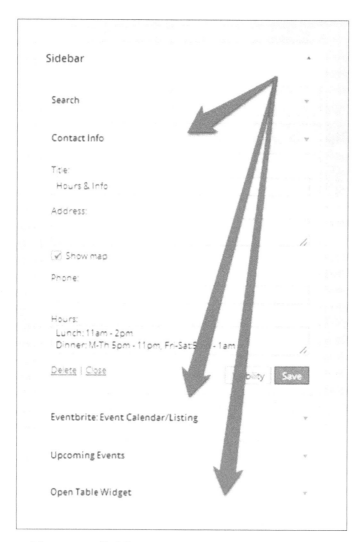

widgets available

NOTES

_____

_____

_____

_____

Website Basics

CHAPTER

3

Getting Started
- 📖 Naming your website
- 📖 Hosting your website

Getting started

Before you can install WordPress, you will need to have a domain name and hosting service. If you already have a registered domain and a hosting service, you should find out from them about adding WordPress to your current site, and read through this chapter before installing. The questions you should ask them include:

- Does your host offer WordPress?
- Can you *add* WordPress to your current site?
- Can you redirect the WordPress URL to your current

NOTES

_____

_____

_____

_____

website?
- Can you just add a page to your current site for your WordPress blog?

We'll answer these questions and more when we go over hosting later in this chapter, but let's start with your name.

Naming Your Website

You have several choices when it comes to the name of your website (domain):
- It *can* be your name or the name of your business
- It *should* be a name that depicts your restaurant niche
- It *should* be short and easy to remember
- It *should* be easy to spell - it will help people find it.
- It *should* be a .com - if someone else already has the .com, and you go with a .net (or something else), guess who will get found first? Most of us just "default" to .com, so keep this in mind when naming your website - especially before you start printing business cards and other marketing materials.

Restaurant
DETAIL

Restaurant Domain Names:
Most restaurants should check for their domain name as they are considering a name for their restaurant. It's how they will be found and remembered..

NOTES

_____

_____

_____

_____

The more specific your website name, the more people will be able to clearly understand what you offer. So if you are Joe's Restaurant, which might be good, but Joe's Steakhouse - joessteakhouse.com is better. It clearly states that you specialize in steaks. And if you want some SEO help, you can also purchase joessteakhouse.com and point it to your joessteakhouse.com website. This can also help you *own* that brand, since you will own all of the ways the domain name might be used.

Trademark

If you create a unique name, or find the "perfect" name, check to make sure it is not trademarked. Before you spend money on logo design, marketing and more, check to see if your name is trademarked. And if you think it is a truly great name, you might want to consider trademarking it yourself.

So, if you have already registered your domain, you can skip this step. If not, here are some articles that can help you understand the importance of a domain name:

- ☐ http://www.sitepoint.com/tips-c
- ☐ http://moz.com/blog/how-to-choose-the-right-domain-name
- ☐ http://www.huffingtonpost.com/tom-lowery/5-tips-for-choosing-a-the_b_3859497.html

NOTES

_____

_____

_____

_____

For help in identifying a domain name:

☐ **Nameboy**
http://www.nameboy.com/
Search for a domain name by inputting keywords or terms that you would like in your domain name. Name-Boy will generate domain names based on the keywords you enter.

☐ **Dot-o-mator**
http://www.dotomator.com/
Search terms (or groups) of terms you would like to use in your domain, and this application searches the avail-ability for you.

You can also search your area and identify keywords that are fre-quently used. These might help you with naming your website. But don't try to be so descriptive that you end up with a very long name or too many names and no real identity. The goal is to find a middle ground that helps identify your territory, while helping brand your business.

Owning Your Name

Once you decide on your domain name, you need to purchase it. Some host companies will offer you a free domain when you pur-chase your hosting from them. Don't do it! A domain name costs under $20/year. If you ever decide to switch your host provider, you may lose that domain.

NOTES

_____

_____

_____

_____

Ownership

As a restauranteur, you understand the importance of ownership, do not "rent" your website or use a "free" address.

So where can you purchase a domain? You can purchase a domain through a registered domain provider which includes some host providers or you can purchase separately and forward to your host. Some things you want to make sure of:

Ensure you have a right to the name. Check for trademark and also check for cybersquatting.

CYBERSQUATTING
domain name registration, trafficking or use of a domain name with the intent to profit from someone else, i.e. trademark - not unlike "squatting" in real estate - it's wrong!

NOTES

_____

_____

_____

_____

- If you are buying a domain name from auction, consult with an attorney to ensure that you have a legal right to the name.

- If a name you want is unavailable, create a "watch" so that if it becomes available, you can purchase it at a later date.

- Check to see if the name you want has been used before and appears in search results. Could this be detrimental to your business?

- Check to see what security features your domain provider offers including:
  - Transfer Lock
  - Name Safe
  - Forwarding
  - Email
  - DNS Management

Hosting your website

If you purchase your domain elsewhere, you need to make sure that your domain "points" to your hosting service. You also need to make sure that your hosting provider offers WordPress hosting. There are hosting companies that offer WordPress packages as well as companies that only offer WordPress hosting.

Some web developers may offer to host your site and manage it for you. The caveat here is to ensure that you own your domain name so that if you ever decide to end that relationship, you can

NOTES

_____

_____

_____

_____

do so without losing your domain name and search traffic you have established. You should also review the contract to understand what your rights are when you end the relationship.

Another factor is to understand what type of control panel your host provides. Many web developers charge extra for using host providers that offer proprietary hosting. This is because there is a learning curve in setting up the website and many times the particular host provider does not offer the support needed.

cPanel
Unix based web hosting control panel that is somewhat of an "industry standard" offered by several host providers

Some hosting tips:
* Check with your web developer to find out what host providers they prefer to work with. They should offer you a variety for you to select from, or they may just recommend that you select one that use the Unix control panel - but it's always a good idea to confirm with them before you purchase hosting service.

* Since you may want to have "landing page" sites for marketing efforts like events, make sure your hosting plan allows

Notes

_____

_____

_____

_____

for unlimited domains and subdomains. You should not need to purchase additional hosting plans for these, but this may mean you need to purchase a more robust plan on some hosts.

- Make sure that your host provider offers the most recent version of PHP. While WordPress will run on PHP 5.2.4 or greater, several frameworks (themes) require 5.4 or greater, and the most recent stable version is 5.5.6.

- Search social media and the internet for feed-back on the host provider. If there are negative comments, research them before you make a decision.

- Make sure that the hosting company offers the minimum requirements set by WordPress:
  - PHP version 5.2.4 or greater
  - MySQL version 5.0 or greater
  - mod_rewrite Apache module
  - *suPHP is an added benefit for security purposes*

Here are some host providers we recommend:
  - ☐ **BLUEHOST**
    http://www.bluehost.com/

  - ☐ **HOSTGATOR**
    http://www.hostgator.com/

  - ☐ **HOSTMONSTER**
    http://www.hostmonster.com/

NOTES

_____

_____

_____

_____

- ☐ **Hub**
  http://www.webhostinghub.com/

- ☐ **JustHost**
  http://www.justhost.com/

- ☐ **Site 5**
  http://www.site5.com/

Managed WordPress host providers

As the title implies, managed WordPress hosting is specifically for WordPress. These hosts usually provide update support and more. However, some of these providers also limit the plugins you can use on your website. So research what you can and can't do before making a selection.

- ☐ **KahunaHost**
  http://kahunahost.com/
  Cloud hosting and Varnish Caching.

- ☐ **Lightningbase**
  http://lightningbase.com/
  Basic management, with per-site backups, caching/CDN setup and more.

- ☐ **Pagely**
  https://pagely.com/
  Pagely handles upgrades, performance, and security for you.

Notes

_____

_____

_____

_____

- ☐ **SYNTHESIS**
  http://websynthesis.com/
  Hosting includes speed and security, as well as content
  marketing and SEO tools

- ☐ **WP ENGINE**
  http://wpengine.com/
  WPEngine emphasizes page load speeds and security.

Shared Hosting vs. Managed Hosting

There are pros and cons to either type of hosting, but here are
some things to keep in mind:

SHARED HOSTING PROS	SHARED HOSTING CONS
Less expensive	Usually older versions of PHP
Easier to use	May not know WordPress

MANAGED HOSTING PROS	MANAGED HOSTING CONS
Dedicated WordPress	Expensive
Performance	May charge per site
Security	Learning curve
Support - WP specific	

Domain Name Redirection

Once you have purchased your domain and hosting, you may
need to do one more thing before you can install WordPress on
your domain, and that is to redirect your domain name to your
host provider. Depending on where you purchased your domain

NOTES

_____

_____

_____

_____

and host, they should provide you with directions or assist you in completing this.

You can redirect a domain name several ways, and here are two:
* Setting nameservers
  Your host company should provide you with two nameservers which are usually in numerical order and you will need to use these to redirect your domain to the host
* A name
  You might want to use this if you are using email from the company that you purchased your domain from, but want to host your site elsewhere. You will need to update your DNS Manager to the IP address provided by the host.

The information you will need from your host company:
* Host provider
* IP address
* Nameservers

You should make sure that you have the following information for future use and/or your web developer:
* Host name
* FTP access
* cPanel access
* Account pin if needed for support

NOTES

_____

_____

_____

_____

Information:
Never email password or access information. Share it via a secure system only.

Once you have your domain name set up on your host, you can install WordPress.

NOTES

_____

_____

_____

_____

Getting Started

CHAPTER

4

Getting Started with WordPress

- 📖 WordPress.com vs. WordPress.org
- 📖 Installing WordPress
- 📖 Understanding the Dashboard

WordPress.com vs. WordPress.org

WordPress offers two different platforms and only one is for self-hosted sites. WordPress.com (not including the VIP program offered by WordPress) offers free hosting for your website as well as several free themes and plugins. However, they are limited to what they offer, and not necessarily what you want or need. Here are some basic comparisons:

	WordPress.com	WordPress.org
Hosting	Free	You need to select provider
Themes	Free – limited	Free and Premium
Plugins	Only those built-in	Free and Premium
Design Modifications	None	Unlimited

NOTES

_____

_____

_____

_____

There are other issues, but I felt that these were the most important differences and strongest reasons to use WordPress.org.

Installing WordPress

There are several ways to install WordPress. You can determine
which way is best for you. Once WordPress is installed, you can
begin modifying the "look" of your site, expanding the functionality of your site with plugins, or adding content.

Before you begin the installation process, you need to make sure
you have the following:

1. Access to your web server

   While you can use your own server including software such as Apache and operating system software
   such as Linux or Unix, this is typically done through
   a hosting service that already provides these items.
   While there are several hosting providers, Word-
   Press recommends four companies (http://word-
   press.org/hosting/).

   Usually the easiest way to access your web server
   is by typing your domain/cpanel or domain:2082
   (domain:2083 if you are using an SSL certificate on
   your site).

2. A text editor

   Some of the files in your installation may need to be
   modified, and to do this you will need a text editor.

NOTES

_____

_____

_____

_____

You will also need a text editor if you want to modify your theme after it is installed.

3. An FTP Client
FTP (File Transfer Protocol) can be used to download or upload files to your site, especially since it is usually much faster than through your browser. The following FTP clients can help you with this.
- ☐ http://filezilla-project.org/
- ☐ http://cyberduck.ch/ (Mac)

Now that you have all of the tools you need, you can install WordPress. There are several different methods of installing WordPress and you can review each one before making a decision on which one you will use. Some of these methods include:

1. Famous 5-minute install
Directions for this process can be found at http://codex.wordpress.org/Installing_WordPress#Detailed_Instructions

2. cPanel
This is done through the hosting provider cPanel. Directions for this process can be found at http://codex.wordpress.org/Installing_WordPress#Using_cPanel

I could go into detail on installing WordPress, but the links above take you directly to the detailed instructions. I recommend either of these installations vs. the quick install and Fantastico installations for several reasons, but mostly for security.

NOTES

_____

_____

_____

_____

WordPress will be installed and you can begin working with your website.

Activating WordPress

Once you have completed the installation process, you will need to activate your WordPress site.

Type your domainname.com/wp-admin in your browser and the following screen will appear:

- Type in your Site Title
- Type in your Username (*use something unique*)
- Type in your password - twice
- Type in your email

NOTES

_____

_____

_____

_____

Welcome

Welcome to the famous five minute WordPress installation process! You may want to browse the ReadMe documentation at your leisure. Otherwise, just fill in the information below and you'll be on your way to using the most extendable and powerful personal publishing platform in the world.

Information needed

Please provide the following information. Don't worry, you can always change these settings later.

Site Title	
Username	
	Usernames can have only alphanumeric characters, spaces, underscores, hyphens, periods and the @ symbol.
Password, twice	
A password will be	
automatically generated for	
you if you leave this blank.	Strength indicator
	Hint: The password should be at least seven characters long. To make it stronger, use upper and lower case letters, numbers and symbols like ! " ? $ % ^ &).
Your E-mail	
	Double-check your email address before continuing.
Privacy	☑ Allow search engines to index this site

Install WordPress

NOTES

_____

_____

_____

_____

The last option is to select whether to have your site visible. But this is an option, and may or may not be searched, regardless of whether you choose not to.

Click Install WordPress.

If you need help selecting a strong password, you can use a site like:

☐ **Strong Password Generator**
http://strongpasswordgenerator.com/
This website generates new passwords and offers password configuration options

☐ **LastPass**
https://lastpass.com
LastPass is a secure password keeper that you can access from your browser, tablet and smartphone.

NOTES

_____

_____

_____

_____

Understanding the Dashboard

Now that you have installed Word-Press, you should start exploring the dashboard and learning what's possible, and to set up your site. And while it seems strange, you are going to start at the bottom of the dashboard with the Settings.

Click on the Settings section and start with General Settings.

GENERAL

The General Settings control the basic functions of your site including title, date and time. While you can change these at any time, you should set them when you first start using your website.

Some of the features you can set here include the time zone you are located in, date format that appears on your site, the email you will use for website notifications and much more. All of these features can be modified at a later date as well, but it is wise to make sure that you set these properly right from the start.

NOTES

_____

_____

_____

_____

NOTE: The screen captures in this book are from the most recent version of WordPress 3.8, and if you have not updated your site, the screens and options may be slightly different.

General

The first section of your Settings is where you can set the date and time of your website. This will effect your website posts, etc.

Note that the default day of the week is Monday. I change this to Sunday, but you can leave it as is.

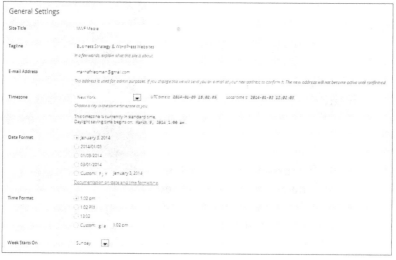

Writing

Formatting

The first option turns emoticons into symbols when checked. The second option will fix invalid XHTML code when checked

NOTES

_____

_____

_____

_____

Default Post Category	The default category for blog posts if you don't assign a category
Default Link	Same as posts, but for your Links
Press This	A "bookmarklet" that can be placed in your browser bookmark bar so you can "grab" bits of content from the web. This feature can help with content creation

Reading

These settings control how your site appears to your readers.

Front Page Displays:

Front Page	If you want your home page to look like a "website" and not a blog, you can create a landing page to appear on your home page and designate another page to show your blog posts (as seen on page 34).
Blog Page Show	You can determine how many blog posts you want on your blog page. The remaining posts will be visible through your archives.
Syndication Feeds	Similar to Blog Pages, you can determine how many posts a reader can see in your RSS Feed
Feed Articles	This determines what is displayed in a feed.
Search Engine	Select if you want to discourage site engines from indexing your site (while setting it up)

NOTES

_____

_____

_____

_____

Reading Settings

Front page displays	○ Your latest posts
	◉ A static page (select below)
	Front page: Home ▾
	Posts page: Blog ▾
Blog pages show at most	10 ◄◊► posts
Syndication feeds show the most recent	10 ◄◊► items
For each article in a feed, show	○ Full text
	◉ Summary
Search Engine Visibility	☐ Discourage search engines from indexing this site
	It is up to search engines to honor this request.

[Save Changes]

Discussion

This is the page that enables you to determine how you want your readers to interact with you when they comment on your site. This is an important part of setting up your website to prevent spamming. There are different points of view on how to handle this. Some people believe that making it difficult to comment on a post will deter people from commenting, so they opt to make it

NOTES

_____

_____

_____

_____

easy, which usually results in a lot of spam. Others deselect most options making it difficult, which creates barriers to discussion. The best solution is the one that works for you, so select carefully.

Other comment settings

Statement 2 - Users must be registered and logged in to comment should be unchecked. Since only registered members can comment, you lessen the likelihood of spam. If you check this box and allow anyone to register before they submit a comment, you increase the opportunity for automated spammers to leave thousands of links in your blog.

Review each of the remaining statements and determine if you want to allow it or not. By selecting the box next to each statement you are agreeing with the action it denotes. A detailed explanation of each comment item can be found in the WordPress codex at http://codex.wordpress.org/Settings_Discussion_Screen.

You should have a spam preventer plugin installed on your site

NOTES

_____

_____

_____

_____

to limit the amount of spam comments that may appear. Your site has Akismet installed, but you will need to activate it. Other spam prevention plugins can be found in the WordPress plugin repository.

Media

Media refers to any items uploaded to your site including images, embeds and files like .pdf.

When you upload an image, it is automatically saved in every size listed unless you have changed the specific file size to 0 by 0.

Image Sizes   You can determine the default sizes for your images.

NOTES

_____

_____

_____

_____

Permalinks

WordPress offers you the ability to customize the appearance of your website permalinks. The shaded URL is an example of the information within each permalink that will appear.

Setting up your permalinks from the beginning is important for SEO, since your pages will be indexed according to their URL. We go into more detail on permalinks for SEO in Chapter 12. But there are some important details you should note when setting up your permalink structure.

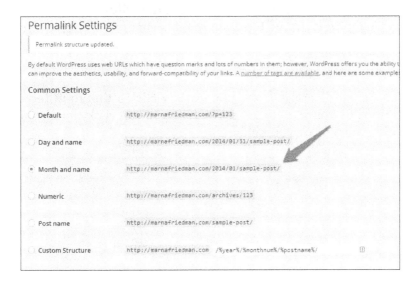

The major permalink items you should be aware of include:

NOTES

- Keep your page and post titles to 3-5 words.
- Leverage your keywords in a descriptive manner when creating titles to make the best use of the three to five words
- Google treats hyphens as spaces between words and underscores as additional parts of words, so use hyphens to separate words instead of underscores

Once the General Settings are complete, you can navigate the other sections of the dashboard.

All of the sections on the Dashboard, except for the menu, can be moved around to customize your view. In addition, by selecting Screen Options (in the top right hand corner of the screen), you can select, and de-select, the items you want displayed on your screen.

And in WordPress 3.8, you can quickly click a link to different sections of your site to get started, or you can click the dismiss option to navigate yourself.

You can click the Help tab, next to Screen Options (see screenshot on next page), to quickly launch help from the WordPress Codex on each screen's topic. The left hand side of the Help screen offers quick links to help on each of the sections displayed on the

NOTES

_____

_____

_____

_____

screen, while the right side provides hyperlinks to further documentation and support forums. The help options change according to the section of the dashboard that is active.

Posts

As you continue to navigate the dashboard, you should explore each section to learn what it offers. This is good to do before you install any themes, since they may add even more sections and functionality to your dashboard.

Posts refer to blog posts that are typically displayed in reverse chronological order, with the most recent first. They can also have categories and tags associated with them. You can filter posts by date or category. And by hovering the post title, you will also be presented with options to edit, quick edit, trash or view the post.

While you can edit a post, one of the nice options is Quick Edit, which enables you to quickly modify some of the features without having to open the post. Items you can change include:

NOTES

_____

_____

_____

_____

- title
- slug
- categories
- add tags
- change the date the post will be published
- deselect the ability to allow comments or pings
- password protect or make a post private
- make a post sticky - it stays at the top of your posts

Posts can be scheduled, private or password protected by modifying the options in the Publish section.

NOTES

In the post editor, you can use Visual mode or Text mode. The visual editor (see above) will only be shown in Visual mode. The box furthest to the right on the top row is called Kitchen Sink, and when clicked will show the second line of options.

Take some time to navigate each symbol to understand what it does and how it can help you.

You can also see that there are unique icons in the top menu of my editor, which are unique to the theme I am using. Plugins and themes can add more icons to your editor for additional features, such as a menu to update selections to a page or post.

The Omega icon offers the ability to use special characters in your pages or posts without formatting changes.

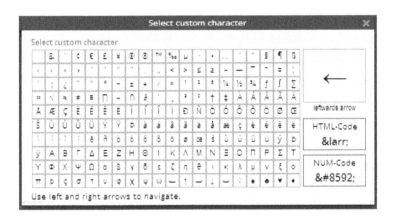

NOTES

_____

_____

_____

_____

Restaurant
DETAIL

When writing posts about your restaurant, you can categorize your posts with the categories of your business, i.e. , location, type of food, price, etc. Tags can add an extra level of detail.

Category: broad based topics to organize the content of your posts, similar to a Table of Contents. Create your category list before you start, since your posts will be indexed and changing them later can be costly, time consuming and basically - a nightmare!

Tag: detailed "keywords" to further organize the content of your posts, similar to an Index

NOTES

_____

_____

_____

_____

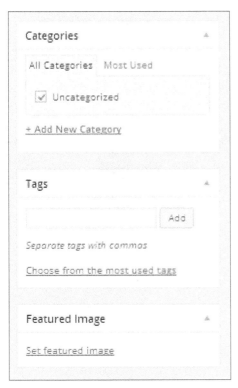

While you may choose keywords for categories and tags, keep in mind that your content should be user focused, and not SEO focused. These terms should relate to the page or post content.

NOTES

_____

_____

_____

_____

_____

Media

Media is where you upload all items to add to your posts and pages.

Most themes will also offer a "Featured Image" which can be found in the lower right corner of your Post dashboard. If it isn't there, make sure "Featured Image" is enabled in the screen options on the top right. You can upload an image to your post which will then be displayed on your Home Page in a "preformatted" size for your theme.

Featured Image

Remove featured image

NOTES

_____

_____

_____

_____

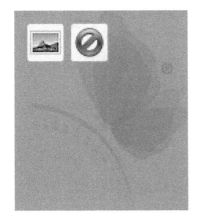

By clicking on the image, you can then select the symbol on the left over the image to edit the image.Click on Advanced Settings for more options.

Size   You can modify the size of the image

Image properties   You can type numbers in the boxes to cre-
                        ate an image border and offset the image
                        from text. The data you enter will populate
                        in the Styles box and you can then go back
                        and edit that by changing the border color
                        and style. You can change the appearance
                        by changing the border style and color
                        (see illustration on page 46).

Visit http://w3schools.com/css/css_border.asp for more information and options.

NOTES

_____

_____

_____

_____

Restaurant
DETAIL

You can link images/logos/pdf files to menu listings and other websites by changing the link URL to their website URL.

NOTES

_____

_____

_____

_____

Borders are a nice way to style food images and differentiate them from other images used on your website.

Pages

Pages are for information that is constant, like an "About" page.

Your restaurant website should include Pages that are relevant to your business such as Menu, Blog, Contact

Comments

As your readers comment on blog posts, you will see a number next to the Comments section of your Dashboard. You can choose to approve, answer, delete or mark the comment as spam.

Appearance

The Appearance section offers you the ability to modify the appearance of your website. Themes will be discussed in the next chapter.

Widgets   Widgets are a function of the theme and can be changed, deleted or added.

Menus   Menus are also a function of the theme and some themes offer multiple menu locations while others offer one. However, you can also create multiple menus as well as custom menus. As you create a menu you can determine if it is the current menu by selecting it in the dropdown on left of your menu Dashboard.

Theme Options is a function of your selected theme as are Background and Header. These will only appear for specific themes.

Editor   You can view your theme files through this option. However, you should not edit any of these files from within WordPress (except maybe CSS) since this can cause your site to "break", and should only be edited through FTP or File Manager. Always backup your site

NOTES

---

before edit any files, and make a copy of the original file before you make any changes.

Plugins

Plugins add functionality to your site. They will be covered later in this book.

Users

This is where all of the users of your website can be found. If you have other people helping you with your restaurant website, you can assign different roles to them which will limit the access they have. While there are plugins that can expand the roles available on your site, here are the default roles:

Administrator   Total access to the site

Editor   This person can write posts and can also manage other people's posts

Author   Can publish and manage their own posts

Contributor   This person can write and manage their posts but cannot publish them

Subscriber   Much like a magazine subscription, this user can only manage their profile

The Dashboard view and access changes with each role.

NOTES

_____

_____

_____

_____

Tools

You can import and export data through the Tools option.

NOTES

_____

_____

_____

_____

WordPress Themes

Learn how to expand the functionality of your WordPress website

- 📖 Understanding Themes
- 📖 Selecting a Theme
- 📖 Installing a Theme
- 📖 Activating a WordPress theme

THEME

The WordPress Theme system is a way to "skin" your weblog. A WordPress Theme is a collection of files that work together to produce a graphical interface with an underlying unifyng design for a weblog.

http://wordpress.org

Definition

NOTES

_____

_____

_____

_____

Selecting the right framework

Themes can provide the look and feel of your website while also offering functionality. WordPress is the platform that the theme works with, it's another component in building your website.

Understanding themes

Now that you have installed your WordPress website, it will help to understand what a theme is. The default theme, Twenty Fourteen, is automatically installed with WordPress and provides some additional functionality for your website including the ability to change the header, background and include several widgets within your site. You can begin to create a list of functions that you would like included in your website. Some of these may be must haves while others are "nice to haves". These lists will help you in your theme selection. A WordPress developer may be able to customize some functionality for you while a graphic designer can custom design your blog and/or website. Some other options include:

☐ **99 Designs**
http://99designs.com/
99 Designs offers crowdsourced designs for websites, logos and more. You host a design contest, create a design brief and select a package. You select the winning designs.

☐ **crowdSPRING**
http://www.crowdspring.com
Offers you the ability to tap into a global pool of creatives offering designs for logos, web, product packaging and design, and much more.

NOTES

_____

_____

_____

_____

If you want to design your own theme, and have it converted to WordPress, you can create a PSD file (Photoshop) and have it converted by a professional service. Some of these services include:

- http://xhtml.pixelcrayons.com/
- http://www.markupbox.com
- https://www.csschopper.com/
- http://www.riaxe.com

Restaurant DETAIL

Remember to consider your restaurant menu functionality when custom designing a theme

Responsive Theme

More than 80% of the world's population now has a mobile phone and 91.4 million of those are smartphones. Research shows that 147.2 million will use tablets by 2015. Given these statistics, you should ensure that your website can be seen and used on multiple platforms. While many WordPress plugins offer the ability to make your theme "mobile friendly", it can effect some of the functionality. But a responsive theme is an approach to web development that enables a website to "respond" to the technology it is being viewed on. A responsive theme quickly adjusts to the new platform and generates a view "fit" for that platform. The view may not be exactly the same across each platform, but provides a better user experience.

A responsive theme is a mobile friendly site, but is not a native app.

NOTES

_____

_____

_____

_____

Selecting a Theme

Making a decision on which theme to use can be overwhelming. This is where a mindmap and/or wireframe can help. Creating a checklist of features can also help (see form on page 56). There are thousands of free themes available on the WordPress repository as well as premium themes from several different theme vendors. The biggest difference between most free themes and premium themes is support. In addition to support, some premium themes include the psd files for you to modify. Premium themes usually include:

- Installation & Upgrade and Forum Support
- Built in SEO and Website Optimization
- Built in functionality instead of using a plugin

Some premium theme vendors include:
- ☐ http://bizzthemes.com/
- ☐ http://colorlabsproject.com/
- ☐ http://www.elegantthemes.com/
- ☐ http://gorillathemes.com/
- ☐ http://www.inkthemes.com/
- ☐ http://ithemes.com/
- ☐ http://www.pagelines.com
- ☐ http://www.studiopress.com/
- ☐ http://templatic.com/
- ☐ http://themeloom.com/
- ☐ http://www.themeskingdom.com/
- ☐ http://themetrust.com/themes
- ☐ http://themify.me/themes
- ☐ http://www.woothemes.com/

In addition to premium theme vendor sites, there are theme

Notes

_____

_____

_____

_____

marketplaces that sell themes from different theme developers. While they typically include support, the support may or may not come from the marketplace. Unlike theme vendor sites, marketplace sites usually include more information like how many times a theme has been sold and user reviews. Some premium theme marketplaces include:

- [ ] http://www.mojo-themes.com/
- [ ] http://themeforest.net/category/wordpress
- [ ] http://wpmarket.net/

Installing a Theme

Once you decide on a theme, you will need to install it on your site. Depending on your theme, you can usually install it from the Themes option in the Appearance section of your WordPress Dashboard.

If you are installing the theme from the WordPress repository, select the options you would like from the list, and review the results. Each of the themes will include a description of what is offered which can help you make your selection. To see a larger version of the theme, just click the Preview hyperlink. Once you make your selection, click the Install hyperlink to install it on your website.

NOTES

_____

_____

_____

_____

Theme Selection

This chart helps determine a theme choice for your website

Theme Name	URL	Page Template	Custom Header	Background	Widgetized Homepage	Responsive	Shortcodes	Slider	Color Options	Custom Sidebars	Restaurant Menu Functionality	Reservations Functionality	Event Functionality	Contact Form	Comments

Installing a premium theme can be as easy as downloading the theme from the premium vendor's website and then uploading to your website. Sometimes the vendor includes additional files in their download, including the .psd files. Read their documentation for specific instructions on how to install the theme.

Once you have installed your theme, you can begin to customize it through your Appearance Dashboard. If your theme includes a Theme Options Dashboard, it will appear in your Dashboard Menu when you install and activate the theme. If you have purchased any premium themes, you will want to remember your user ID and password to access their Support forums. You should also make sure to keep a record of your receipt numbers for future reference.

NOTES

_____

_____

_____

_____

You can install multiple themes, but you can only activate one at a time. You can preview a theme on your site after you install it by clicking on the live preview button. Once you activate a theme, you can begin customizing it.

Activating a Theme

If after activating the theme, you decide to change it, you can just upload another one and activate that. Some things to keep in mind when you switch themes:

- any customization you did on the current theme will be gone when you activate the new theme - it will still be in your database, but not visible on the new theme, i.e., menus
- you will need to activate any custom menus you created and assign a location
- you will need to upload logos and favicons
- shortcodes will probably be different, so you will need to

NOTES

_____

_____

_____

_____

update any you created on the previous theme
- page templates will be different, so your content may look different
- you will probably discover other differences, just be sure to remember any customizations you created, so you can update them with a new theme.

NOTES

_____

_____

_____

_____

NOTES

_____

_____

_____

_____

WordPress Plugins

CHAPTER
6

Learn how to expand the functionality of your WordPress website

- 📖 Understanding plugins
- 📖 Selecting plugins
- 📖 Installing plugins

Definition

Plugins

Plugins are tools to extend the functionality of WordPress. Plugins offer custom functions and features so that each user can tailor the site to their specific needs.

- http://wordpress.org

NOTES

_____

_____

_____

_____

Plugins can quickly expand the functionality of your WordPress restaurant business website. These plugins can enhance the user experience, expedite menu search, integrate social media and much more.

Understanding plugins

Now that you have installed your WordPress website, it will help to understand what a plugin is. WordPress is the platform you will use, but much like a theme, plugins are another building block. While your theme expands the design of your site, plugins expand the functionality. As we discussed in Chapter 4, some of this functionality may be built into your premium theme, while plugins may offer this where it doesn't already exist.

There are specific functions that should be included in any site, but there are several plugins that offer each of these functions. You need to make an informed decision on these.

Selecting Plugins

How do you determine the best plugin for your site? Like the chart on page 56, you should create a chart for each of the plugin categories that can help you make an informed decision. As you navigate the different plugins available in each category, you will be able to identify the features that are important to you and your readers. While you may want certain features, are they important to your visitors, or important to you? The answer to this question should always be your visitors.

Key plugin functions for any site are:

- SPAM preventer

NOTES

_____

_____

_____

_____

- SEO
- Backup
- Analytics
- Contact form
- Page navigation
- Cache
- Editorial Calendar

How do you decide which plugin?

The WordPress plugin repository provides an opportunity for a lot of information that can help you make your decision. But you need to understand how to use this information.

- Does the plugin offer the functionality you need?
- Is the plugin compatible with the current version of WP?
- How many times has it been downloaded?
- When was it last updated?
- What version of WordPress does it require?
- What version of WordPress is it compatible to?
- Is the developer supporting the plugin?
- What are the forum comments?
- Does it conflict with other plugins you are using?
- Are there screenshots?
- Is there documentation?

Another consideration would be if the plugin is a free version of a premium plugin. Typically this is indicative of a developer that is supporting the plugin, at least the premium plugin. You can test the free version and upgrade to the premium when you are ready.

NOTES

_____

_____

_____

_____

Installing Plugins

Once you have selected a plugin, you can click on the Install Now button. Since this installation can affect your website, WordPress asks you to confirm this decision. Once you complete the installation, you will still need to activate the plugin.

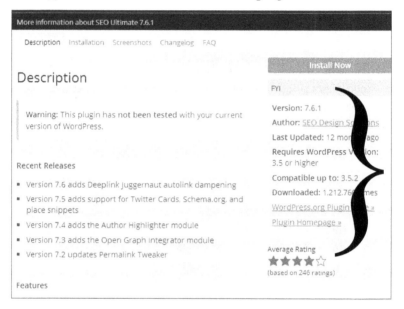

Plugin musts:

- You should be sure you have access to FTP or your host service whenever you install a plugin, in case it "breaks" your site and you need to uninstall it.

- Always deactivate a plugin before deleting it. And check to see if the developer has added any other details you need to be aware of if you delete the plugin.

NOTES

_____

_____

_____

_____

- Plugins can conflict with other plugins and themes. Keep track of when and which plugins you have installed. If you do have a conflict, you typically have to deactivate all of your plugins and then reactivate one by one to identify the conflict.

- A plugin that has been working fine may conflict with a new plugin, so usually the only way you can identify an issue is to deactivate and activate one by one.

Basic Plugins for Every Website

Earlier in this chapter, we discussed key plugins that every website should have. Now we will suggest different plugins from the WordPress repository for each of those functions.

SPAM PREVENTER

WordPress comes with Akismet preinstalled, and this is a great SPAM preventer. You will need to have a key to activate this plugin.

Some other SPAM preventer plugins include:

- ☐ Anti-SPAM by Cleantalk
- ☐ Anti-SPAM Bee
- ☐ SPAM Free WordPress

NOTES

_____

_____

_____

_____

SEO

Search Engine Optimization (SEO) is an important aspect of every website and requires much more than a plugin. But a plugin can help you set your site up properly for SEO.

- ☐ All in One SEO Pack
- ☐ Greg's High Performance SEO
- ☐ WordPress SEO by Yoast

ANALYTICS

Analyzing how visitors are using your website can help you improve their experience and increase your impact. You can do this with Google Analytics, and we will review how to set this up later in this book, but you can also use plugins on your website to help you with this including adding your analytics to your WordPress dashboard.

- ☐ Clicky Analytics
- ☐ Google Analytics Dashboard for WP
- ☐ WP Slim Stat

BACKUP

We will go into detail about backups in a later chapter, but in the meantime, here are some backup plugins that are available on the WordPress repository for you to use.

- ☐ BackWP Up
- ☐ UpdraftPlus - WordPress Backup and Restoration
- ☐ WP-DB-Backup

NOTES

_____

_____

_____

_____

Contact Form

A contact form is a great way to capture emails and contact information for future follow-up. This is one of the areas where I recommend a premium plugin, but there are several free contact form plugins available also.

- [ ] Contact Form 7
- [ ] Contact Form Maker
- [ ] Visual Form Builder

Policy

You should include a Privacy Policy and Terms of Use on your website. You can include these as pages on your website and link to them in your footer, or use a plugin. Since these documents pertain to legal issues, it is **advisable to consult a professional**.

- [ ] Auto Terms of Service and Privacy Policy
- [ ] Agreement

Cache

There are lots of different things that can slow your website down. A cache plugin can help you improve the speed of your website, which in turn can help the user experience and SEO results.

- [ ] Quick Cache
- [ ] W3 Total Cache
- [ ] WP Super Cache

Notes

_____

_____

_____

_____

SOCIAL MEDIA INTEGRATION

There are several social media integration methods, and using a plugin is one of them. Here are some plugins that will help you integrate your social media on your website.

FACEBOOK
- ☐ Add Link to Facebook
- ☐ Flexo Facebook Manager
- ☐ Simple Facebook Connect

GOOGLE PLUS
- ☐ Google Plus Google
- ☐ Google Plug Widget

LINKEDIN
- ☐ LinkedIn Share Button

PINTEREST
- ☐ Pinterest Badge
- ☐ Pinterest Pin It Button
- ☐ Pretty Pinterest Pins

TWITTER
- ☐ Twitter Badge Widget
- ☐ Tweet Blender
- ☐ Xhanch MY Twitter

Some plugins include additional settings and configuration in order to use them. If so, the plugin may note that with a Set-

NOTES

_____

_____

_____

_____

tings option. You should review the plugin documentation before moving forward to ensure that you are installing the plugin correctly.

Other plugins require additional resources in order to function properly, which may incur an additional cost. This information is usually in the plugin documentation, but if you did not review that before installing, you may not be aware until you try to activate the settings and complete the installation.

oEMBED

While there are several YouTube plugins available, you should also understand that WordPress offers a feature called oEmbed. This feature enables you to paste a URL from specific sites into your WordPress site including the ability to automatically paste a video or image onto your post or page. You do not need a plugin to use the oEmbed function.

For security purposes, you can only embed URLs matching an internal whitelist. So if a URL is "private", you will not be able to embed it.

NOTES

_____

_____

_____

_____

How is this done?
1. Go to the application (see list below)
2. Find the image/video you want
3. Copy the URL from the browser
4. Go back to WordPress page/post
5. Ensure you are in HTML mode
6. Paste URL
7. Save

Websites that offer the oEmbed functionality include:
- [ ] blip.tv
- [ ] DailyMotion
- [ ] Flickr (both videos and images)
- [ ] FunnyOrDie.com
- [ ] Hulu
- [ ] Instagram
- [ ] Photobucket
- [ ] PollDaddy
- [ ] Qik
- [ ] Revision3
- [ ] Screenr
- [ ] Scribd
- [ ] SlideShare
- [ ] SmugMug
- [ ] SoundCloud
- [ ] Twitter
- [ ] Viddler
- [ ] Vimeo
- [ ] WordPress.tv (only VideoPress-type videos for the time being)
- [ ] YouTube (only public videos and playlists - "unlisted" and "private" videos will not embed)

NOTES

_____

_____

_____

_____

The oEmbed Process

1. Type your content/post and then in text mode
2. Select the item you want to include from one of the websites listed on page 70
3. Copy the URL

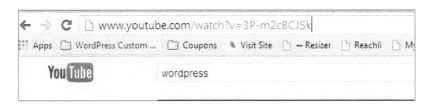

4. Paste the URL into your page or post - make sure you are in text mode

5. That's it. You can preview to see if it worked correctly. If it didn't, confirm that you pasted the URL onto your website while you were in text (HTML) mode and not visual.

NOTES

NOTES

_____

_____

_____

_____

Plugins for Restaurants

Learn how to expand the functionality of your WordPress website
- 📖 Types of plugins for restaurants
- 📖 Selecting plugins

We've seen plugins that can expand the functionality of a basic WordPress website, and there are also plugins for menus and events that can quickly convert a basic WordPress website into a restaurant and/or catering business website. These plugins can be added to your website to provide more information and engagement for your visitors.

Types of plugins for restaurants
There are so many plugins available that it is easy to overload

NOTES

_____

_____

_____

_____

your site with them which can result in a slower site, possible code conflicts between plugins, and ultimately confusion for your visitors. So review the information and make your selections wisely.

Here are some items you may want to include in your website:

RESTAURANT MENU PLUGINS
As a restaurant, one of the most important functions on your website will be offering menus. If you have selected a theme that does not offer this, then you will need a plugin to add this functionality to your website.

- ☐ Easy Restaurant Menu Manager
- ☐ Eewee Restaurant Menu
- ☐ Food and Drink Menu
- ☐ Locu for Restaurant Menus and Merchant Price Lists Plugin
- ☐ Restaurant Menu Manager

DAILY SPECIALS
One way to keep patrons coming back to your website might be to offer daily specials. A plugin can help you add this functionality to your website.

- ☐ DeMomentSomTres Restaurant
- ☐ Easy Restaurant Menu Manager
- ☐ JRWDEV Daily Specials

NOTES

_____

_____

_____

_____

RECIPES

While you certainly don't want to share any of your "trade" secrets, you might want to periodically share some of your unique recipes.

- [ ] All In One Schema.org Rich Snippets
- [ ] Easy Recipe
- [ ] GetMeCooking Recipe Template
- [ ] WP Ultimate Recipe

OPERATIONS

It is important for your visitors to know your restaurant hours, and a plugin can quickly add this functionality to your website.

- [ ] Business Hours
- [ ] Custom Business Locations
- [ ] Local Search SEO Contact Page

EVENTS

Many restaurants offer special events for holidays, and other occasions. With the help of a plugin, you can add this functionality to your restaurant website.

- [ ] All in One Event Calendar
- [ ] The Events Calendar
- [ ] WordPress Event Calendar

EDITORIAL CALENDAR

Blogging has become a very hot topic for restaurants, and other websites. Sharing information with your community can increase your reputation, and the regular addition of

NOTES

_____

_____

_____

_____

content to your website can impact your search engine results. An editorial calendar can help you manage this process, especially if you have someone else blogging besides you.

- ☐ Betaout Content Cloud
- ☐ Edit Flow
- ☐ Editorial Calendar

Selecting Plugins

Review each of the categories to determine which functionality you would like to add to your website. Prioritize the plugins
so that as you install them, if there is a conflict, you will know
which ones you really want to have. Another factor to consider
is how easy the plugin is to use and integrate with your theme. If
the plugin requires a significant amount of customization, then
you may want to reconsider it.

Some ther factors to consider:
- Do you really need it?
- Is this functionality available in your theme?
- Does the plugin have a premium option if you choose to
  upgrade and for support?
- Will you use it?
- Is the plugin being supported?

NOTES

_____

_____

_____

_____

Reservations & Online Orders

Offering your visitors more options on your WordPress website
- 📖 Accepting & Managing Reservations
- 📖 Accepting Online Orders

Expand your website by offering online reservations, ordering and delivery functionality with additional plugins and/or HTML code.

Accepting & Managing Reservations

There are plugins available that can help you integrate automated reservation functionality onto your website, making it easier for customers to reserve tables through an automated process while aincreasing your productivity.

NOTES

_____

_____

_____

_____

Here are some items you may want to include in your website:

RESERVATION PLUGINS
As a restaurant, you may want to include the ability for your visitors to make reservations. If you have selected a theme that does not offer this, then you will need a plugin to add this functionality to your website,

- ☐ easyReservations
- ☐ ReDi Reservations
- ☐ Tablebooker

Or if you are using a premium online reservation system like Open Table, then you can use a WordPress plugin to add this to your website, or use HTML code provided by the company to place in a text widget (more about this in Chapter 10).

Accepting Online Orders

You are probably already accepting credit cards at your restaurant, so you can check with them about adding the ability to accept online orders for pickup and delivery. Some of these plugins may require a premium service in addition to a merchant solution in order to be used.

ONLINE ORDER PLUGINS
- ☐ EasyWay Online Ordering
- ☐ WPMenuMaker
- ☐ Zuppler Online Ordering

NOTES

_____

_____

_____

_____

Page Builder

Learn how to create page layouts for your WordPress website
- 📖 Understanding Page Templates
- 📖 Using Page Builder Plugins

Understanding Page Templates

WordPress offers page templates that can change the "appearance" or layout of different pages on your website. The default page template in WordPress is page.php. Your Theme may also include other custom page templates, such as pages for:

- Home
- Team
- Portfolio
- Blog
- Contact

NOTES

_____

_____

_____

_____

- As you explore different themes for your website, determine what pages you might need and if those page templates are included. If not, you would have to create them, or possibly use shortcode options. But there is another option, and that is using a page builder.

Page Builders have been around for a while, but have become increasingly popular as WordPress users seek to expand the functionality of your website.

If you choose to create your own page templates, refer to the WordPress codex at http://codex.wordpress.org/Page_Templates.

Using Page Builder Plugins

Several premium themes now include page builder options, but if they do not, there are free page builder plugins available on the WordPress repository as well as premium plugins. The plugins on the WordPress repository include:

- ☐ Aqua Page Builder
- ☐ IG PageBuilder
- ☐ MotoPress
- ☐ Page Builder by SiteOrigin

Page Builder plugins can be activated on any page and once activated, you use them instead of the WordPress editor. They enable you to quickly create unique page layouts, and once created, can often be saved as "templates" to reuse the same layout on similar pages, i.e. , services.

NOTES

_____

_____

_____

_____

Understanding the pros and cons are important as you begin your website development.

Pros	**Cons**
quick page layouts	too much can be overwhelming
templates without code	may not be maintained
see what's possible	layouts may differ in themes

Look at the features offered in the different page builders to see if the functionality you are looking for is included. If not, move on to another one until you find the one that includes what you are looking for. You can always have custom page templates created based on the pages you create using a page builder, as long as the functionality is included in your theme or you have your website custom coded, i.e. , toggles, tabs, etc.

Restaurant
DETAIL

You can create page templates for catering packages that offer consistency and quick formatting on your website.

Most of the page builder plugins are drag and drop offering several options. Some of them may conflict with your theme and other plugins, so you should test them to ensure that all of the functions are working.

NOTES

_____

_____

_____

_____

Premium Plugins

As we discussed, there are also premium page builder plugins that you can purchase to use on your site. Since these are premium plugins, they typically include support, which might be a key factor in your decision.

☐ **Visual Composer by WP Bakery**
http://vc.wpbakery.com/

☐ **Elegant Builder by Elegant Themes**
http://www.elegantthemes.com/gallery/elegant-builder/

☐ **Builder**
http://themify.me/builder

☐ **Dynamik Genesis Extender**
http://cobaltapps.com/downloads/genesis-extender-plugin/

☐ **Ether Content Builder WordPress Plugin**
http://ether-wp.com/ether-builder/

NOTES

_____

_____

_____

_____

Learn how to expand the functionality of your WordPress website

- 📖 Using Widgets
- 📖 Creating Widgets
- 📖 Selecting widgets

Sometimes you don't need to use a plugin on your website, but you can use a widget instead.

Using Widgets

There are so many widgets available for you to use on your website, and there are pros and cons to doing this. While adding HTML code to a text widget can be easier than uploading a plugin that might conflict with your current site, it might be even more

NOTES

_____

_____

_____

_____

of a risk.

Keep in mind that according to the terms of use on many of these widgets, you cannot modify them. And while it is great that other websites make these widgets available for free, they are actually causing you to incur a cost, which is the cost of lower SERP.

So why would I even include this information? Because you need to know, and make informed decisions about what to include on your website. Adding too many plugins to your website can create issues with other plugins and also slow your site down.

Creating a Widget

1. Select the widget you would like to use.
2. Go to the website and create the widget according to their directions.
3. Copy the HTML code that they give you.
4. Go back to your WordPress website Dashboard > Appearance > Widgets.
5. Select a text widget and drag it to the sidebar or footer location you would like the widget.
6. Paste the HTML code in the widget, add a title and save.
7. If you want to paste the widget on a page instead, then:
8. Go to the page where you want the code.
9. Select Text option in your editor.
10. Paste the code.
11. Save the page.

NOTES

_____

_____

_____

_____

Selecting Widgets

Now that we have explained the pros and cons of using widgets from the larger syndicated real estate websites, here are some resources for you to find widgets that you might want to use on your website.

Online Reservations

☐ **Eveve**
http://www.eveve.com/restaurants.htm
Offers a widget to integrate their premium system.

☐ **Free Bookings**
http://www.freebookings.com/en
The basic functionality of this program is free, but they also offer premium upgrades like table management at an additional cost.

☐ **Open Table**
https://www.otrestaurant.com/marketing/Reservation-Widget
This widget requires that you have an account with Open Table for reservations.

☐ **Reservation Genie**
http://www.reservationgenie.com/
Offers an economical option for online reservations and table management.

NOTES

_____

_____

_____

_____

Event Information

☐ **EventBrite**
http://developer.eventbrite.com/doc/widgets/
Add event signup widgets to your website for upcoming
special events without the need for a plugin

NOTES

_____

_____

_____

_____

Google Tools

Learn how to leverage Google tools to monitor and analyze your WordPress website
- Google Webmaster Tools
- Google Analytics
- Google Places
- Google Plus
- Google Authorship
- Google Hangouts
- Google Keyword Planner
- Google Alerts

Google offers a suite of tools that can help you with your website. This chapter provides an overview of the tools and a basic idea of what each one provides to help you manage your website.

NOTES

_____

_____

_____

_____

Google Webmaster Tools

Google Webmaster Tools monitors your site for website performance including indexed pages and crawl status along with site traffic. Much like other Google tools, GWT is about metrics: what is indexed, what is linked, and what is getting traffic. Since this is a Google tool, it does not include results from other search engines.

The tools offered through Google Webmaster Tools include:
- Submit and check a sitemap
- Generate and check a robots.txt file, and helps to discover pages that are blocked in robots.txt by chance.
- List internal and external pages that link to the site
- Get a list of broken links for the site
- See what keyword searches on Google led to the site being listed in the SERPs, and the click through rates of such listings
- View statistics about how Google indexes the site, and if it found any errors while doing it
- Designate a location

Monitoring your results will help you see which pages have higher CTR (click through rate) and then determine some strategies to improve their rankings. You also may discover that your visitors are interested in topics that you weren't aware of, which will help you in your content strategy.

NOTES

_____

_____

_____

_____

Once you have set up your Google Webmaster Tools account, you will need to add your site to your account and then verify your website, which can be done fairly quickly if you are using a plugin like WordPress SEO by Yoast.

Webmaster Tools

You can use the boxes below to verify with the different Webmaster Tools, if your site is already verified, you can just forget about these. Enter the verify meta values for:

Google Webmaster Tools:

Bing Webmaster Tools:

Alexa Verification ID:

Google Analytics

Google Analytics, like GWT, shows total traffic for a website. As SERP becomes more sophisticated, you will see more long-tail search terms which means instead of "joe's pizza in austin tx", you might see "joe's pizza in four points, austin tx" as clicks to your site. Google Analytics may tell you how you were found, which can also help you in your content creation strategy.

As you become comfortable reviewing your analytics, you can expand your strategy based on the results. Create a spreadsheet of keywords that you want to be found with, along with the terms you *are* being found with. Develop blog posts around these terms and monitor the results. You should also leverage this information in creating your local SEO strategy.

NOTES

_____

_____

_____

_____

Analyze your competition. Review the websites and social media presence of your local competition by doing a keyword search and seeing who comes up in the results. You can increase your results by sharing relevant content and adding new content on a consistent basis.

You will need to create a profile for your website on Google Analytics in order to measure your results.

Once you create the account profile for your website, you will need the tracking information to input on your website.

1. Click on the Admin link on the top right side of your screen. Select Tracking Info from the middle screen.
2. Click Tracking code.
3. Copy the code.
4. Your theme options panel may include a location for this

NOTES

_____

_____

_____

_____

code. If it doesn't, you can use a plugin like Insert Headers and Footers.

5. You will also need to connect your Google Analytics and Google Webmaster Tools. Go back to Property Settings in Google Analytics and scroll down to Google Webmaster Tools.

Webmaster Tools Settings

Webmaster Tools site optional ?

If your property is also a verified website in Webmaster Tools, and you are the owner, you can associate your Webmaster Tools data here. Google Analytics will then be able to display some of that data in some reports.

Google Places

https://www.google.com/business/placesforbusiness/

As local search becomes more important, it is even more important to confirm your business location with Google through Google Places to be found on Google Search, Maps, Google+, and mobile devices. Since more than 90% of the public searches for businesses online, this is critical to your visibility and SERP. Once your are verified with Google Places, your clients will be able to post Google Reviews for your business, which is a metric factor in SEO.

Even though you have completed the process of signing up your business with Google Places, you will not be verified until you receive a postcard from Google to verify your business. Once you receive the postcard, you need to input the code on your Google Places profile to verify your business, and then your business should begin to appear in local search results.

NOTES

_____

_____

_____

_____

Google Plus

http://www.google.com/+/learnmore/

Create a business page on Google Plus and start sharing information and engaging with other members of the community. Don't make the mistake of excluding this platform from your social media strategy, it is growing rapidly and through its features such as Google Hangouts, is increasing in adoption. You also need a Google Plus profile in order to activate Google Authorship, which brings us to the next Google tool you should have.

Google Authorship

https://plus.google.com/authorship

This is a fairly new Google Tool and only requires your activation. Not only does Google Authorship put a face to your content, but when you establish yourself as the original author of the content, Google ranks it higher for you than duplicate content. Your claiming ownership as the author increases its value.

Google Authorship is all about you.

Google has started to implement Author Rank which ranks Google Authorship writers based on their engagement and social factors. Early adoption of this program increases your visibility which in turn can impact your Author Rank.

NOTES

_____

_____

_____

_____

Google Hangouts

http://www.google.com/+/learnmore/hangouts/

Through your Google Plus profile, you can create and invite people to Google Hangouts. These are online gatherings where everyone invited can take part in the conversation. You can have live calls with clients, schedule a neighborhood workshop on events at your restaurant, collaborate with your team. Google Hangouts works on smartphones, tablets and desktops, and it's even accessible through Gmail.

Google Keyword Planner

https://adwords.google.com/ko/KeywordPlanner/Home

The Google Keyword Tool has been replaced with Google Keyword Planner. The new tool is a little more difficult since previously you could get broad match statistics for keywords, with Keyword Planner, you'll get historical statistics only for exact match.

You can search for keywords based on relevant terms and/or landing pages. Get historical statistics and traffic estimates to determine if there is value in using the terms you have selected. You can review the estimated clicks to get an idea of how successful your strategy and/or campaign might be. And if you are considering Google Ads, these estimates can also help you determine a campaign budget.

The best strategy would be to create a local SEO campaign that focuses on relevant long-tail keywords.
• Using zip codes in your content, but still keeping it longtail,

NOTES

_____

_____

_____

_____

so 'Italian restaurant in Five Points, Fort Lauderdale Beach 33316' and incorporating this into your content. But don't make it so specific that it is beyond what someone might search for.

- Optimize your business directory listings with the local search terms you are focusing on. Be creative in how you include them.
- Review your social media profiles and optimize them with these keywords as well.
- Review your Google Webmaster Tools and Analytics results to understand what your visitors may be looking for and adjust your content and social media strategy to provide that and increase engagement.

Google Alerts
http://www.google.com/alerts
Google Alerts can be helpful in online reputation management as well as content generation. You can create Alerts that let you know what is being said about you. You can also create Alerts to keep apprised of what's being said about topics relevant to your niche.

As you create Alerts, you can determine what you want to know:
- frequency
- type of results
- how many
- where to send results

NOTES

_____

_____

_____

_____

Using WordPress SEO by Yoast

There are several SEO plugins that you can use on your Word-Press website, but I use WordPress SEO by Yoast, so the following pages will take you through my basic set-up of this plugin. These are the settings *I use* and yours might be different. I strongly recommend that you review the videos and tutorials on the plugin site to familiarize yourself with the options available and make the decision that fits your situation. This information is to illustrate how I use the plugin, and not to dictate how you should use it. *Please review the developer's tutorials.*

1. If you are currently using a different SEO plugin, then you should transfer the data before you deactivate the plugin. And yes, you will need another plugin to do this. SEO Data Transporter can help you with this.
2. Once you complete that step, deactivate the old plugin and delete it so you can install WordPress SEO by Yoast.
3. Go to Settings on WordPress SEO by Yoast. You can click on the tour button if you want to explore the different options of the plugin, or watch the videos offered. You can customize these once you are more familiar with the setup. Again, these screens are just an example of how I use this plugin.

NOTES

_____

_____

_____

_____

4. The next screen will be Titles & Meta. There are several tabs that you will need to go through to set up your site. So start with General Settings and move through each tab.

GENERAL

Title settings - the first option is to Force rewrite titles. But this is auto detected and will show whether it needs to be done or note, so I leave it alone.

SITEWIDE META SETTINGS

One of the important aspects of setting this plugin up is to try and prevent duplicate content from being indexed.So I check the option to **Noindex subpages of archives**. *If you blog a lot, your archive pages will increase and the only page that is important to index is the first one, so checking this will help prevent the additional pages from being indexed.*

The only other thing I check here is **Use meta keyword tags**. There is mixed feedback on this option, but I typically check it. You can choose to select this or not.

Those are the only items that I find important to select on the General Settings tab.

5. Moving on to the Home tab where I set up my home page template. While I do use defaults on this page, I modify what I use. The purpose of using the variables is so that if you ever update the site title or tagline, it will be automatically updated and you won't have to change it manually. Some people convert these variables to numbers, creating a Understanding the variables may help you determine what to choose.

NOTES

_____

_____

_____

_____

General | Home | Post Types | Taxonomies | Other

Title settings

☐ Force rewrite titles

WordPress SEO has auto-detected whether it needs to force rewrite the titles for your pages, if you think it's wrong and you know what you're doing you can change the setting here.

Site wide meta settings

☑ Noindex subpages of archives

If you want to prevent /page/2/ and further of any archive to show up in the search results, enable this.

☑ Use meta keywords tag?

I don't know why you'd want to use meta keywords, but if you want to, check this box.

☐ Add noodp meta robots tag sitewide

Prevents search engines from using the DMOZ description for pages from this site in the search results.

☐ Add noydir meta robots tag sitewide

Prevents search engines from using the Yahoo! directory description for pages from this site in the search results.

%%sitename%% %%page%% %%sep%% %%sitedesc%%

%%sitename%% - your sitename

%%page%% - since this is your homepage template, it won't have a page, so you can delete this.

NOTES

_____

_____

_____

_____

%%sep%% - is a separator, which is usually a dash, but since this is typically determined by your theme, it might be some other type of separator. Since you might change your theme, this could change as well. My choice is to delete the variable and actually use my own separator which is the pipe (|) a vertical line. It's your choice.

%%sitedesc%% - site tagline or description in your Word-Press General Settings.

So my final choice is:
%%sitename%% | %%sitedesc%%

6. I type my site description in the next box and since I selected meta keywords in General Settings, you will see the meta keywords box for me to type keywords in.

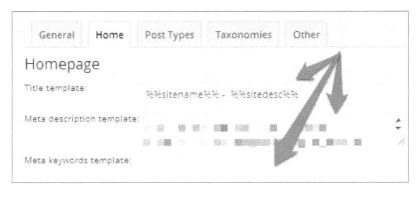

If you choose to write your own title and description, just type a number in the boxes as you see on page 99.

NOTES

_____

_____

_____

_____

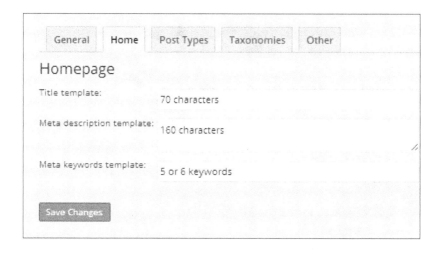

The next tab has several sections to set up the default Title, Description and Keywords for posts, pages and media. Depending on your sitename, you may just want to use the default %%sitename%% . If your sitename is not very keyword focused or "geolocal", you may want to use the character option here and create a unique title for every page. The caveat here is that you will need to remember to type in a title for every page and/or post you create. The default includes %%page%% which you can remove.

Since I do want my posts to be indexed, I make sure to leave the box for Meta Robots **unchecked**. Here are my final post options.

NOTES

_____

_____

_____

_____

NOTE: if you want to use the focus keyword option in the WordPress SEO meta box on the bottom of your post pages, then you can type %%focuskw%% in the Meta keywords template, and this will include this in your search results. Some people agree with this practice, others don't, so again, make your own decision.

7. Pages will be similar to posts, but some may choose to have their pages set to no index, and then just set options on specific pages through the meta box that appears at the bottom of each page and change the settings on that individual page. Only you can determine the Meta Robots option for your pages, since you know how you set up your pages and the content they contain. I leave it alone.

NOTES

_____

_____

_____

_____

8. Media refers to your images, pdf uploads, etc. which I typically do not want indexed, so the Meta Robots option will be checked in this section.

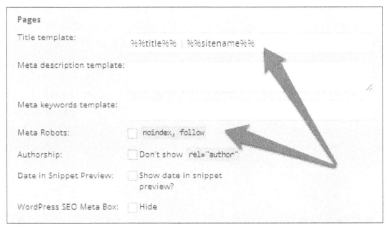

9. If you are using plugins that offer a custom post type, like events or something else, it will probably be listed on this screen. These are usually an enhancement to your posts or pages and not mandatory to your SERP.

Remember that anything you check to noindex, nofollow will not be indexed or ranked by Google. This process can be overwhelming, so you might want to get some help with this that is specific to your website and how you have set it up, both with the IDX and the website content.

NOTES

_____

_____

_____

_____

10. Since on my website, the Taxonomies would be duplicate content, I select Meta Robots to noindex, follow. The content has already been indexed in its original format, so I don't want it indexed again. An exception to this would be if you design your category pages to include an introduction about the category and then just an excerpt from each page or post.

To set up the latter option, you would need to create a category template within your website page templates. You should also include a description of each Category and Tag on your website.

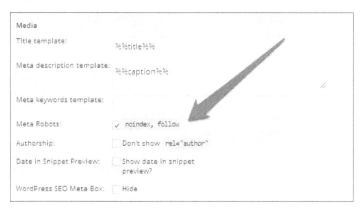

You can modify the Title templates as you see fit for your website similar to how I did above. And notice that I included description options for both Categories and Tags. If you do not, then leave the description field blank.

NOTES

_____

_____

_____

_____

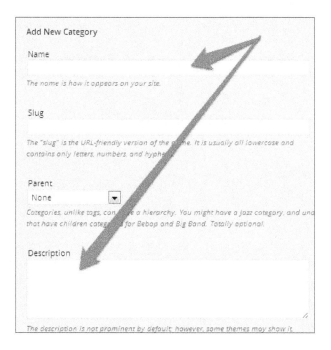

11. The last tab is for the meta data on your pages and posts which I pretty much leave as is.

NOTE: If your author name is the name of your site, then delete the sitename option from your author. Otherwise, it will show up in search results as Jane Doe Jane Doe.

NOTES

_____

_____

_____

_____

Categories

Title template:

%%term_title%% Archives %%page%% %%sep%% %%siter

Meta description template:

Meta Robots: ☐ noindex, follow

WordPress SEO Meta Box: ☐ Hide

Tags

Title template:

%%term_title%% Archives %%page%% %%sep%% %%siter

Meta description template:

Meta Robots: ☑ noindex, follow

WordPress SEO Meta Box: ☑ Hide

Format

Title template:

%%term_title%% Archives %%page%% %%sep%% %%siter

Meta description template:

Meta Robots: ☑ noindex, follow

WordPress SEO Meta Box: ☑ Hide

NOTES

_____

_____

_____

_____

| General | Home | Post Types | Taxonomies | Other |

Author Archives

Title template:

%%name%%, Author at %%sitename%%

Meta description template:

Meta Robots:
☐ noindex, follow

☐ Disable the author archives

If you're running a one author blog, the author archive will always look exactly the same as your homepage. And even though you may not link to it, others might, to do you harm. Disabling them here will make sure any link to those archives will be 301 redirected to the homepage.

Date Archives

Title template:

%%date%% %%page%% | %%sitename%%

Meta description template:

Meta Robots:
☑ noindex, follow

12. The next section is Social, where you can integrate your website with your social media.

In Facebook, these settings relate to your business page. Although you will have to connect through your personal Facebook account, this is just to approve access to your page. Here are the settings I recommend:

NOTES

_____

_____

_____

_____

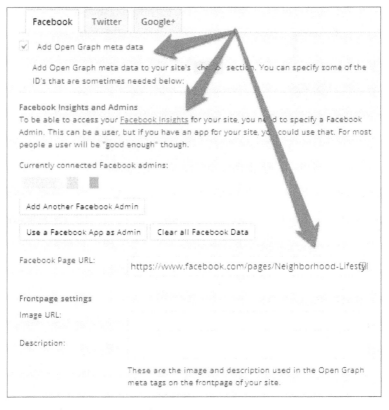

Here are the Twitter settings I use:

NOTES

_____

_____

_____

_____

Google:

Enter your author name, or select from the dropdown. If you have created a Google+ page for your business enter the URL here, and then make sure to go to your Google+ business page and enter your website's URL in the about section.

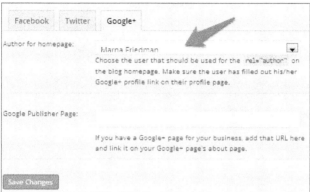

The next section is your sitemap. Here is where you can select the options that appear in your sitemap to be indexed. If you selected no index, follow on categories and tags, then you can select to not include them in your sitemap as well.

NOTES

_____

_____

_____

_____

XML Sitemap

☑ Check this box to enable XML sitemap functionality.

You can find your XML Sitemap here: XML Sitemap

You do not need to generate the XM_____n, nor will it take up time to generate after publishing a post.

User sitemap

☐ Disable author/user sitemap

General settings

After content publication, the plugin a_____tically pings Google and Bing, do you need it to ping other search engines too? If s_____the box:

☑ Ping Yahoo!
☑ Ping Ask.com

Exclude post types

Please check the appropriate box below if t_____'s a post type that you do NOT want to include in your sitemap:

☐ Posts (post)
☐ Pages (page)
☑ Media (attachment)
☐ Agents (myteam)

Exclude taxonomies

Please check the appropriate box below if there's a taxonomy that you do NOT want to include in your sitemap:

13. Permalinks

    The first option is a personal preference. Some people like category in the search results, because they think it helps the search engines to know where they are. The next section refers to Stopwords, which are common words sometimes not considered by search engines to increase search speed.

NOTES

_____

_____

_____

_____

There are arguments for checking and unchecking. It's your decision. I do recommend selecting Redirect attachment URLs to parent post URL so that attachments will automatically redirect to the post or page where it was attached.

Permalink Settings

☐ Strip the category base (usually /category/) from the category URL.
☐ Enforce a trailing slash on all category and tag URL's

If you choose a permalink for your posts with `.html` or anything else, like a / on the end, this will force WordPress to add a trailing slash to post pages nonetheless.

☐ Remove stop words from slugs.

This helps you to create cleaner URLs by automatically removing the stopwords from them.

☑ Redirect attachment URL's to parent post URL.

Attachments to posts are stored in the database as posts, this means they're accessible under their own URL's if you do not redirect them, enabling this will redirect them to the post they were attached to.

☐ Remove the `?replytocom` variables.

This prevents threaded replies from working when the user has JavaScript disabled, but on a large site can mean a huge improvement in crawl efficiency for search engines when you have a lot of comments.

☐ Redirect ugly URL's to clean permalinks. (Not recommended in many cases!)

People make mistakes in their links towards you sometimes, or unwanted parameters are added to the end of your URLs, this allows you to redirect them all away. Please note that while this is a feature that is actively maintained, it is known to break several plugins, and should for that reason be the first feature you disable when you encounter issues after installing this plugin.

Canonical Settings

Force Transport:
Leave default

14. Breadcrumb Settings
Since most themes offer breadcrumb settings options, I usually leave this page blank and defer to the theme settings.

NOTES

_____

_____

_____

_____

15. RSS
The only change I make on this page is to reword the way I want my content linked in RSS posts. There are explanations of each of the defaults to help you in your selection.

16. Import & Export
This is where you would import from another plugin, and as you can see, the recommendation is to use the plugin we mentioned on page 92 first.

17. Edit files
Unless you know what you are doing, you should not edit these files.

That's it! Keep in mind that SEO is not a set-it and forget-it program. You will need to continue to monitor your search results and make adjustments or additions as necessary.

The best thing you can do is to blog. Adding fresh, relevant content to your website will reflect activity to the search engines and they will want to know what is going on.

NOTES

_____

_____

_____

_____

Integrating Social Media

Learn how to integrate social media on your restaurant website.
- 📖 Growing Your Community
- 📖 Integrating Social Media Platforms

Social Media

Social media is an online conversation betweenreal people. It's not about the technology, it's about the relationships you will create and build.

Definition

NOTES

_____

_____

_____

_____

Growing your community

Even though your website gives you an online presence, you need to augment it with social media. Not every social media application is right for your audience. Understanding how to use social media will help you to engage with your community.

It's important to understand how social media can help with your business strategy. Before you begin to use social media, you need to register for the applications you want to use. Even though you may not use all of the applications listed below, you may want to register for an account to use in the future. If your user name is consistent, it is easier for your community to find you. A quick way to find out if the name you want is available is to use namechk. This application will do a quick search of the name you select and let you know if it is available.

Namechk
http://namechk.com/

NOTES

_____

_____

_____

_____

You can type your username in the top box and click OK. Keep in mind that you should use a name that is less than 20 characters, especially since tweets from Twitter are only 140 characters. It's also easier to remember a shorter name. The screen will quickly change to show you which sites have that username available, which names are taken, and also if there are domains with that username available. Clicking on each site will take you to the site, where you can register.

As you select the sites and register, be sure to remember your user ID and password.

According to Nielsen, almost 80% of the US internet users frequent social networks and blogs. And close to 40% of social media users access their sites on smartphones - which is another reason why your website needs to be responsive.

Using Social Media

Most restaurant websites include links to their social media on their marketing materials and websites. But they aren't using the social media. Much like a conversation, if you aren't participating, you really aren't involved. And social media is not about your restaurant business, it's about your community. You need

NOTES

_____

_____

_____

_____

to be involved with the conversation to engage current and future customers.

Social media is about engagement and not sales. Participation in social media offers an opportunity to engage your community helping to create and build relationships. Conversations on social media are indicative of what people want to talk about.

Start a conversation and wait for participation. Always remember that this is a conversation, **not a sales pitch**. Sometimes it's great to just listen too. You can learn what people want to talk about.

You can also create networking events so that your online community can meet in person. The earlier you start your networking events, the more opportunity you have to grow your community and increase your business. Invite your clients to participate in the community. They become a representation of your ability to build relationships.

Unlike many other businesses, restaurants often have social media representation without their being involved. Sites like Yelp may offer reviews and customer interactions without a restaurant even being aware, especially if they aren't active on social media.

As you can see from the infographic on page 115 from Balihoo, local marketing and participation is an opportunity for all local businesses.

http://balihoo.com/

NOTES

_____

_____

_____

_____

THE LOCAL WEB
THE SINGLE LARGEST OPPORTUNITY FOR NATIONAL BRANDS

The Local Web is a growing ecosystem of online media channels collectively driving local awareness & sales

73% of online search activity is related to local content

Group deal use up **33%** by SMB between June & December 2011

7 in 10 consumers are more likely to use a local business if it has information available on a social media site

LOCAL SEARCH — Yahoo! Google Bing

DAILY DEALS

In June 2011 foursquare reported hitting **10** million users

Gartner predicts mobile phones will overtake PCs as the most common web access device worldwide by 2013

LOCATION BASED

59% of all local-business searchers agree that ratings & reviews are important while searching for a business

LOCAL WEBSITES (THE HUB FOR ALL LOCAL MARKETING)

MOBILE — LOCAL REVIEWS

LOCAL MARKETING MATTERS
A transformational shift is underway as new media channels & shifting consumer behavior change the way brands market

BENEFITS OF LOCALIZED MARKETING

67% GREATER CUSTOMER RELEVANCE, RESPONSE & RETURN

82% OF LOCAL INTERNET SEARCHES FOLLOW-UP OFFLINE VIA IN-STORE VISIT, PHONE CALL OR PURCHASE

39% BETTER CONVERSATION & CONNECTIVITY AMONG CUSTOMERS

THE # OF MEDIA SOURCES USED BY CONSUMERS **5.2** IN 2010 TO **10.4** IN 2011 (100% INCREASE IN 12 MONTHS)

29% IMPROVED LOYALTY & ADVOCACY

IN 2011, MORE THAN $1.1 TRILLION IN RETAIL SALES COULD BE ATTRIBUTED TO WHAT IS REFERRED TO AS "WEB-INFLUENCED" PURCHASES
DEFINED BY FORRESTER AS OFFLINE RETAIL SALES THAT ARE INFLUENCED BY ONLINE RESEARCH

27% BRAND DIFFERENTIATION, DISTINCTION & PREFERENCE

49% OF MARKETERS BELIEVE LOCALIZED MARKETING IS ESSENTIAL TO BUSINESS GROWTH

DIGITAL MEDIA IS EXPECTED TO REPRESENT **23.6%** OF ADVERTISING BY **2015**

BUILD YOUR LOCAL MARKETING PLAN

STEP 1	STEP 2	STEP 3	STEP 4
STRATEGY	LOCAL WEBSITES	TACTICS	MEASUREMENT
Determine role of local marketing in national strategy	Develop local websites; automated content capability is an advantage	Use local web media channels to reach consumers locally & drive traffic to local websites & dealers	Use local marketing to learn, modify & increase marketing ROI

Balihoo
Local Marketing. Automated.

Sources: 1 CMO Council 2011 Localize to Optimize Sales Channel Effectiveness, 2 TMP/ComScore 2009, 3 BIA Kelsey, 4 Google/Shopper Sciences, Zero Moment of Truth Macro Study Industry Studies, US, April 2011, 5 GroupM Search with Kantar Media Compete: From Intent To In-Store: Search's Role In The New Retail Shopper Profile, Oct 2011, 6 Google, 7 TMP/IS Miles, 8 Gartner 2010, 9 comScore Networks/TMP Directional Marketing, 10 Merchant Circle's 7th SMB Survey, Dec 2011

NOTES

_____

_____

_____

_____

Using Twitter

Twitter presents an opportunity for quick, interactive conversations. A quick explanation of Twitter is: when you search for something on the internet, you can find what exists. But when you search that topic on Twitter, you hear what people are saying about it. Restaurants should want to be part of the conversation. You also want to monitor what is being said about you, and/or your restaurant and/or catering business. And there are tools that can help you with this. Twitter is available on your desktop, smartphone and tablet.

Name: This should be your name or name of your business.

Location: Geo-location is becoming more critical. Note that you can always edit your profile.

Website: This should be your website so that people know where to go to learn more about you and/or your restaurant.

Bio: Be informative, you only have 160 characters. This should not be a sales pitch.

Facebook: You can elect to post your Tweets to Facebook. Keep in mind that not everyone uses Twitter. So sharing your Tweets on Facebook helps to engage that community as well.

Remember to save changes.

When you sign up for Twitter, you should input the following information:

Photo: Upload an image that relates to you and/or your profile. Don't leave this blank, build your brand.

NOTES

_____

_____

_____

_____

Profile

This information appears on your public profile, search results, and beyond.

Now edit your photos and bio right from your profile.   ✕

Photo

Change photo ▾

This photo is your identity on Twitter appears with your Tweets.

Header

Change header ▾

Recommended dimensions of 1252×626
Maximum file size of 5MB
Need help? Learn more.

Name

Marna Friedman

Enter your real name, so people you know recognize you.

Location

Atlanta, GA

Where in the world are you?

Website

http://mpressivesolutions.com

Have a homepage or a blog? Put the address here.

Bio

Real Estate Agent in Dallas & Acworth, GA at
http://marnafriedman.com; Business Strategy &

About yourself in 160 characters or less.   38

Facebook

f Connect to Facebook

Post Tweets to your Facebook profile or page.

NOTES

_____

_____

_____

_____

To create a Twitter widget for your website, you can click on the Widget option on the left side of your Twitter profile.

1. Click on the gear at the top of your Twitter screen.

2. Select Edit Profile from the dropdown. And then select Widgets from the left sidebar.

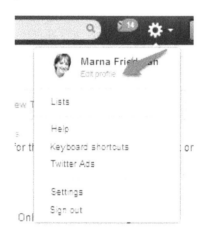
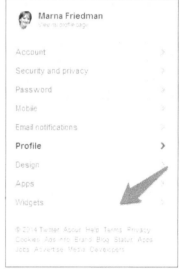

NOTES

_____

_____

_____

_____

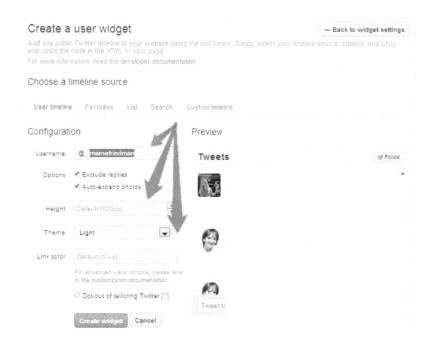

3. Make sure to complete all the relevant information as you create your Twitter widget as noted above. You can modify the size and color to coordinate with your website. And the widget will look like the Preview shown above. Once you have completed all of the details, click the Create widget button to capture the HTML code.

NOTES

_____

_____

_____

_____

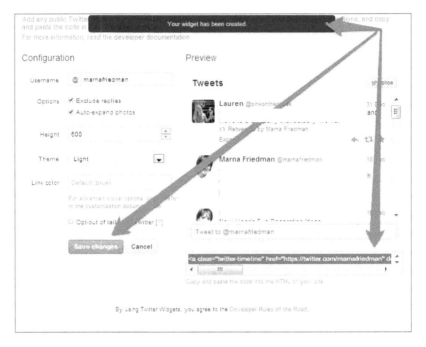

4. Copy the HTML code in the bottom box to paste into a text widget on your website.

5. Refer to Chapter 10 for instructions on how to use Widgets. Then place a text widget where you want your Twitter widget to appear and paste the HTML code into the widget. Save. View your page, as you can see from the example on page 122.

NOTES

_____

_____

_____

_____

Text ▼

Title:

Chat with us on Twitter

<a class="twitter-timeline"
href="https://twitter.com/marnafriedman" data-widget-
id="_____ _____ _____ ____">Tweets by
@marnafriedman</a>

tE men

n.t itter
);}

☐ Automatically add paragraphs

Delete | Close

Save

NOTES

_____

_____

_____

_____

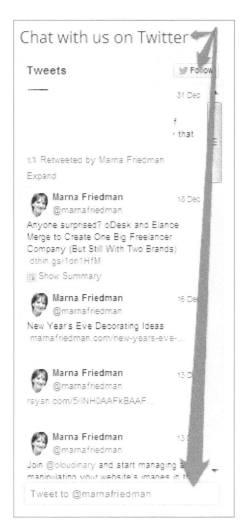

The benefit of using the Twitter widget is that you will not have to worry about updating a plugin or code conflicting. It is also responsive on a responsive website. You also will not need all of the OAuth information you probably need for a plugin.

The visitor can elect to Follow you on Twitter, as well as participate in the Twitter stream right from within your website, so engagement is easy.

If you would like to use a plugin instead, there is a list of some that you can use on the next page.

NOTES

_____

_____

_____

_____

Twitter Plugins
- ☐ My Twitter Widget
- ☐ Twiget Twitter Widget
- ☐ Twitter

Using Facebook

While Twitter is about the world, Facebook is about friends and family. As you share information, you will continue to grow your Facebook community. But you need to determine if your community is on Facebook. You can also advertise through Facebooks Ads, and if your community is active on Facebook, a Facebook page might provide a decent ROI.

Some of the questions you need to answer are:
- Are your customers on Facebook?
- Can you engage visitors in a conversation?
- Does Facebook offer a community for your business?
- Will your fan base grow based on content?

You also have the ability to add 13 apps to your Facebook Page Timeline, so you can select apps that are relevant to your business.

Other ways you can leverage Facebook for your restaurant and/or catering business include:
1. Sharing information about upcoming neighborhood events.
2. Share some menu item information
3. Answer questions, address concerns, and acknowledge when people take the time to post on your Page.
4. Update your page with news items and engage friends to ask what they are interested in.

NOTES

_____

_____

_____

_____

Facebook offers WordPress plugins in addition to those that are available from the WordPress repository. You can read about the options from Facebook at https://developers.facebook.com/blog/post/2012/06/12/facebook-integration-for-wordpress/.

Facebook Plugins
☐ Facebook Social Plugin Widgets
☐ Facebook Widget
☐ ThemeLoom Widgets

Using Pinterest

Pinterest is one of the fastest growing social media platforms. According to comScore, Pinterest was the fastest independent site to hit 10 million monthly unique visitors in the U.S. As participation in Pinterest continues to climb, businesses are also quickly discovering ways to connect with individuals. Pinterest is all about visual sharing.

Let your community know that you are on Pinterest and to share photos by integrating it onto your website. Share photos from menu items, events you have participated in around town, anything to engage current and future customers.

So how do you use Pinterest?
- Think visual
- Organize your boards - consider your audience
- Be descriptive so your visitors are compelled to repin
- Use relevant keywords to be found easily
- Engage, don't sell

To join Pinterest, you just need to register at the Pinterest web-

NOTES

_____

_____

_____

_____

site. In order to create a widget, you will need to join Pinterest as a Business.

1. Go to https://www.pinterest.com/business/create/
2. Complete the information requested, and click Create Account.

The first time you login to Pinterest, you should familiarize yourself with what is possible. Click on the About button in the upper right corner, and navigate through all of the sections.

General Information About Pinterest
- Organize your boards by interests, with one interest per board - this makes it easier for viewers to follow
- Update your boards regularly - so only create as many as you can maintain
- Use a call-to-action on your pins to engage others
- Add text to images, make your pins descriptive and targeted
- Use searchable keywords in board names
- Select a compelling image for your board cover

Like Twitter, Pinterest offers the ability to create widgets for you to add to text widgets on your WordPress website with a little modification of the code provided.

1. Click your profile name in top right corner of Pinterest. Scroll down to Visit Help Center and click to learn what's possible.

NOTES

_____

_____

_____

_____

2. Type WordPress Widget in the search box.

3. You will have three options to select from and should explore all three.

NOTES

_____

_____

_____

_____

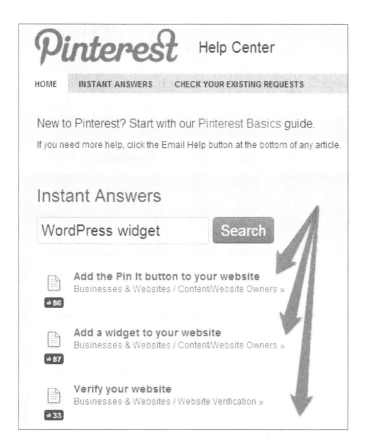

The third option is to verify your website is probably where you should start. Once you verify your website, you will also have access to analytics, which can help you in learning how your Pinterest activity is doing.

Pinterest offers several different ways to verify your website, as well as directions to walk you through the process with

NOTES

_____

_____

_____

_____

several different host providers. Complete this step before you move forward with creating a widget.

4. The next step is create your widget. Go to http://business. pinterest.com/widget-builder/#do_pin_it_button and select the widget you want on your website. For this example, I am selecting profile widget.

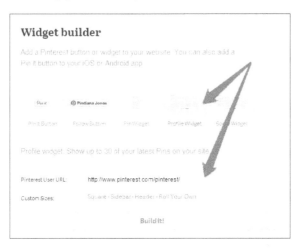

5. Click on your Pinterest profile on the right hand side of your screen to copy your URL.

NOTES

_____

_____

_____

_____

6. Paste the URL in the Pinterest URL box and select your widget option. You can modify size later.

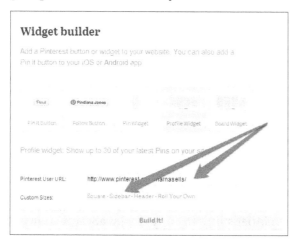

You can modify the dimensions if you like. And then click Build it!

7. Below your preview, you will see code. You can either go back and choose Roll Your Own, or modify the size within the code - but only change the measurements!

8. You will actually need to modify the top code a little here since the javascript provided will not work in a text widget, and you will need the link code as well. But if you click the help link at the bottom of the page, it will direct you to a page where you can copy code. See the steps and final code on page 136.

NOTES

_____

_____

_____

_____

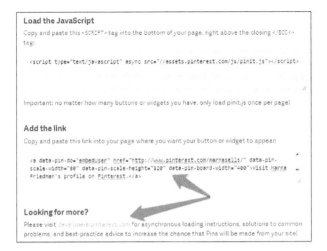

Load the JavaScript

Copy and paste this <SCRIPT> tag into the bottom of your page, right above the closing </BODY>
tag:

```
<script type="text/javascript" async src="//assets.pinterest.com/js/pinit.js"></script>
```

Important: no matter how many buttons or widgets you have, only load pinit.js once per page!

Add the link

Copy and paste this link into your page where you want your button or widget to appear:

```
<a data-pin-do="embedUser" href="http://www.pinterest.com/marnasells/" data-pin-
scale-width="80" data-pin-scale-height="320" data-pin-board-width="400">Visit Marna
Friedman's profile on Pinterest.</a>
```

Looking for more?

Please visit developers.pinterest.com for asynchronous loading instructions, solutions to common
problems, and best-practice advice to increase the chance that Pins will be made from your site!

9. Go to the Developer's link to copy the JavaScript code you
   will need.

If you'd like to load the JavaScript asynchronously, you can do so by copying and
pasting this snippet anywhere on your page:

```
<script type="text/javascript">
(function(d){
    var f = d.getElementsByTagName('SCRIPT')[0], p = d.createElement('SCRIPT');
    p.type = 'text/javascript';
    p.async = true;
    p.src = '//assets.pinterest.com/js/pinit.js';
    f.parentNode.insertBefore(p, f);
}(document));
</script>
```

You only need to load this script once per page, no matter how many buttons or
widgets you use.

NOTES

10. Copy the code, being sure to use it only once per page! Add the Link code below that code and copy all of that code into your text widget where you want it to appear as below.

11. If you want to feature individual boards, you can create a widget for each one. Add a description line before the link code. Paste link code for each board, but only use JavaScript code once. Save the widget and view.

NOTES

_____

_____

_____

_____

Join us on Pinterest

Some samples from our Boards

 Marna Friedman

See On Pinterest

The great thing about using the Pinterest widget is that if you have created boards for different niche markets, you can include the board on those landing pages of your website. Then of course, there are also the widgets that Pinterest offers for you to try, as well as WordPress plugins to integrate Pinterest on your WordPress website.

NOTES

_____

_____

_____

_____

Pinterest Plugins
- ☐ Pinterest Master
- ☐ Pinterest Pinboard Widget
- ☐ WordPress Canvas Widgets

Of all of our inventions for mass communication, pictures still speak the most universally understood language.

- Walt Disney

Using YouTube and Flickr

Videos and photos also add another visual component to your business. You can create a YouTube page for your website and share the videos from your business there. This can also link back to your website, Facebook, Pinterest, LinkedIn and you can Tweet about it.

Ways to use videos and photos:
- Add videos to website
- Share tidbits of information for your community
- Create videos of menu items, events, and more.
- Interview customers about their dining experience

To incorporate YouTube and Flickr on pages and widgets, you

NOTES

_____

_____

_____

_____

can use the oEmbed function explained in chapter 5 on page 69.

For a widget of your Flickr photos though, you will need a plugin.

Flickr Plugins
- ☐ Fast Flickr Widget
- ☐ Flickr-stream
- ☐ Slickr Flickr

Using LinkedIn

LinkedIn is a valuable tool for connecting with like-minded professionals. You can quickly create a group on LinkedIn and begin to have discussions on relevant topics. Building a community on LinkedIn offers an opportunity to share information and ask questions of decision-makers who in turn can support your business.

As a professional group on LinkedIn, you are also interacting as an individual instead of a brand, making it more personal. Prepare some relevant discussions before launching your group. Also participate in other groups that are relevant to your brand and business. Always keep in mind that you should be focused on sharing information and building a community, not about selling. While you are participating in other groups, learn who the key influencers are and invite them to connect with you.

Review your current network and determine if some of your connections can introduce you to others. Understanding how to leverage the community you already have will help in continuing

NOTES

_____

_____

_____

_____

to help it grow.

Join local business groups and network with other people in your area. This can be a great way to share information and build lead generation. LinkedIn has significant opportunities to gain exposure for restaurant and/or catering businesses as a location for networking and possible customer event opportunities.

Like Twitter and Pinterest, LinkedIn also offers a widget that you can integrate onto your website. You can choose to show or hide connections. To create a LinkedIn widget for your website, go to https://developer.linkedin.com/plugins/member-profile-plugin-generator.

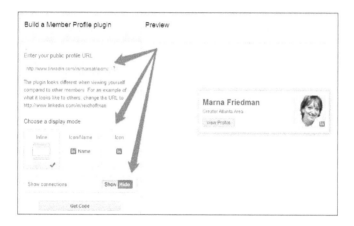

And you can see the preview as you select options. Click Get Code and copy the code to paste into a text widget to place in your selected location on your website.

NOTES

Notes

_____

_____

_____

_____

Location Based Applications

❦ CHAPTER ❦
13

Learn how to leverage your location
- 📖 What is location based technology
- 📖 Using location based apps

Navigating location based technology
Understanding location based technologies that are available and what they offer will help you decide which ones to use and how to integrate them into your restaurant's website and social media strategy.

NOTES

_____

_____

_____

_____

What is location based technology

Location based apps (LBS) utilize GPS (Global Positioning System) information.

Maintained by the US government, GPS is a space-based satellite navigation system that provides location and time information, anywhere on earth, where there is an onstructed line of sight to four or more GPS satellies. This technology is freely accessible by anyone with a GPS received.

Definition

This technology has come together to offer location-based social networking. Leveraging this technology can help to grow your community. Location based technology can be used to:

*   recommend social events in a city
*   requesting the nearest business or service, such as an ATM or restaurant
*   turn by turn navigation to any address
*   locating people on a map displayed on the mobile phone
*   location-based mobile advertising
*   connecting with other people nearby
*   games where your location is part of the game play, for example your movements during your day make your avatar move in the game or your position unlocks content.

NOTES

_____

_____

_____

_____

Using Location Based Apps

Location based apps help to engage your customers in social networking and can help to grow your community. One application, Poynt, states "the free personal concierge that turns you into a local expert, wherever you are." Foursquare offers more than "check-ins", it offers a game component including badges and mayorships; while also offering information on services near your location.

☐ **Foursquare**
https://foursquare.com
Foursquare offers free services to businesses including claiming their venue and offering specials. And once you claim your venue, you will also have access to analytics including:
- Total daily check-ins over time
- Your most recent visitors
- Your most frequent visitors
- Gender breakdown of your customers
- What time of day people check in
- Portion of your venue's Foursquare check-ins that are broadcast to Twitter and Facebook

As visitors check-in to your venue, they may also be provided with some suggestions nearby. For out-of-town visitors, this can be a great way to connect.

NOTES

_____

_____

_____

_____

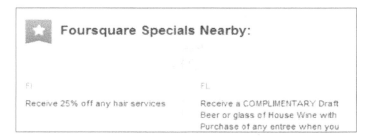

You can claim your business on Foursquare here- http://business.foursquare.com/listing/. Once you have claimed your business, you can add it to your website as a social media link or add a text widget to display more information.

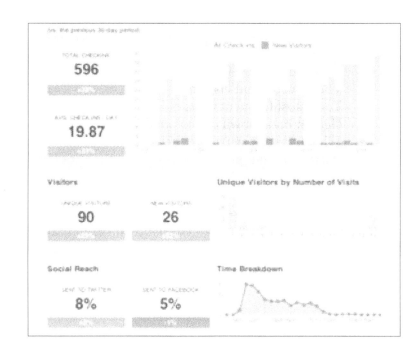

NOTES

_____

_____

_____

_____

This data can be helpful in planning future events and marketing your restaurant. Looking at the analytics also helps you see the use of the application and integration with other social media platforms like Twitter and Facebook.

Other "check-in" apps are available, but offer limited services compared to Foursquare. However, if these applications are used by your customers, then you should investigate them.

Location based apps are an evolving technology and applications like Foursquare are surviving by understanding how people are using it and offering services to respond to this. Other applications are coming into play and as usage increases, it will become clearer how LBS is evolving and responding to different types of content generation and aggregtion. The analytics being offered by Foursquare provides a window into this by leverage the data to understand consumer behavior.

Location based apps for restaurants

The following are apps that can be helpful for restaurants and are available on iPhone and Android, and you can add the icons to your social media profile for your website.

☐ **Poynt**
http://www.poynt.com/

☐ **Yelp**
http://www.yelp.com/

☐ **YPLocal**
http://www.yellowpages.com/

NOTES

_____

_____

_____

_____

Notes

_____

_____

_____

_____

Social Share

Learn how you can extend your message beyond your website and engage your community
- 📖 Sharing content
- 📖 Engagement
- 📖 Getting them back to your website

If you haven't heard it yet, here it is - if you want to become known as an expert and grow your business, you need to share information and engage your visitors. The easiest way to do this is through blogging.

Sharing content
Many businesses, even restaurants, think that having a Facebook page and posting new menu items and specials is social sharing.

NOTES

_____

_____

_____

_____

_____

Let me tell you now that this is wrong, in so many ways. First of all, blasting about your menu is not sharing. And the only people that may like your post are friends. But if you create a blog post about an amazing new menu selection and how you are doing something unique, you can not only share the post on Facebook, but you now shared something that visitors might find interesting. The post is incidental, the story is about what you did that was different and how other customers might like it or, even better, be interested in dining at your restaurant.

Before you create great content, you need to have all of your tools in place. So having share buttons on your posts is important. The easier you make it for people to do something, the more likely they are to do it.

How to incorporate social share

Your theme may include social share buttons. If it does, I recommend you use them since adding a plugin when one isn't necessary can slow your site down. But if you still want one, then here are some options.

- ☐ Cunjo: The Best Free Social Share Plugin
- ☐ Hupso Share Buttons for Twitter, Facebook & Google+
- ☐ Social Login, Social Sharing, Social Commenting and more!

You will need to go through the set-up process and determine which platforms you want included in your share options. There are so many social media platforms and adding all of the options will overwhelm your visitors. But not including some options may limit sharing of your content. This is where you need to con-

NOTES

_____

_____

_____

_____

sider your audience, and your own search results.

Some of the share plugins offer analytics which might be help-ful in seeing which posts create engagement. But here again, you need to consider what you can manage. All of the analytics can be overwhelming and divert your attention. If you are creating relevant content and sharing it, the engagement will happen.

Some of the plugins will offer share buttons with counts in them reflecting how many times a post was shared on a particular plat-form. This is great for large sites, but probably not for a niche site. So you should select icons that reflect the platform, but without numbers.

Another item you might be able to customize is where your social icons appear. There are several places where this can be including at the top of your post under your title or at the bottom when the article ends just before comments. And then there are widgets that might appear on the right or left hand side of your website. I prefer my social share buttons at the bottom of my blog posts. It's a preference. Is sharing easy and intuitive for them?

I like the large callout to Share before the buttons on Hupso, as well as the option to email or print a post. And I limit the share platforms to the five above because I believe those are the ones my visitors use. If I was in a different industry, perhaps the list

NOTES

_____

_____

_____

_____

of sites would be different because my visitors use different platforms.

How you share

Now that you have the share buttons for your visitors to share, you need to have a plan so that you can share your content. As you develop your website content, you will want people to read it. So the more exposure you can get for your content, the better the opportunity for it to be seen. Sometimes this can be just another task to get done, unless you automate it.

Adding an autoposting plugin to your WordPress website will automate the process of sharing your content. Some of them will be more involved to set-up, so be prepared. Keep in mind that the additional installation steps are for your protection. I recommend that you read through all of the documentation and review the screenshots to help you make a decision.

Jetpack by WordPress.com offers a module on autopublish among its many different features. This might be a good place to start, since you can also enable some other functionality with the same plugin including site analytics on your dashboard.

Here are some autopublish plugins you might use:

- ☐ Jetpack by WordPress.com
- ☐ NextScripts: Social Networks Auto-Poster
- ☐ Social Media Auto Publish

Once the plugin is set-up, it will autopublish your posts to the social media platforms you have selected. This all sounds great

NOTES

_____

_____

_____

_____

right?

Tips on sharing strategies[1]:
1. Posts with 80 characters or less get 60% more engagement.
2. Question posts get 100% more comments.
3. Photos get 53% more Likes, 104% more comments, and 84% more click-throughs

Timing
Understanding when people use social media can be an important part of your content and sharing strategy. Think about where your visitors are, or who you want your visitors to be. If they are like most people, then they use social media during their breaks from work, before or after work. If you post at 10a, then you are probably not reaching them. And since WordPress offers the ability to schedule your posts, which in turn would schedule your sharing, then you need to be strategic in when this occurs.

Tips on timing[1]:
1. Focus on **when** you post: Facebook peak activity is at 3pm
2. Facebook activity peaks on Wednesday
3. Posting 1-2 a day gets 40% more engagement
4. Posting 1-4 times per week gets 71% more engagement

[1]http://blog.kissmetrics.com/more-likes-on-facebook/

While the statistics above give you a basic idea of when and what to post, these are statistics, and may not be relevant to your community. Be strategic in the timing of your posts to see if it impacts your engagement.

NOTES

Engagement

The most important part of sharing is to engage with your visitors. The timing of your posts and your sharing may be consistent with the statistics, or unique to you. The only way you will learn this is by testing. Test the timing of your posts, test the way you create your message, and also look at how you engage. What gets you to click on a link? What gets you to share a post?

Another strategy would be to look at your competition. Are they blogging? What is their engagement? What content is relevant to readers? Visit other sites and see what content they are sharing. Is it relevant to your niche?

Things that can effect engagement
- Photos create interest and offers an opportunity for your content to be virally shareable
- Regular blogging increases engagement
- Provide meaningful content

Getting them back to your website

All of the sharing is great but the objective is to get them back to your website. How many times have you read a great article and then visited the author's website to see what else they have shared? That's who you want to be. A visitor is probably not going to scroll through your Facebook posts. But they might look at other articles on your website. The organization of your content needs to be easy for them to navigate.

While your content is being autopublished to social media sites,

NOTES

_____

_____

_____

_____

you should engage in Facebook and Google groups participating in discussions. Find groups that are relevant to your niche and contribute to the conversation. Share information beyond what you have posted in social media sites. Offer reasons for them to visit your website. But remember to put the megaphone away. Build a reputation as an expert in your niche, become a resource. And don't forget to listen. Listen to the other conversations that are going on. What do people want to know and how you can help them. These groups can not only help you with engagement, but can also help you with content ideas.

As you share information on your website, redirect visitors there. If you wrote an article about decorative displays for food items that featured a special product, don't send visitors to the product website, send them to your article. Sometimes you become so intent on sharing, that you forget that little detail.

NOTES

_____

_____

_____

_____

Notes

_____

_____

_____

_____

Email

Learn how to expand the functionality of your WordPress website
- 📖 Email Plugin
- 📖 Creating a List
- 📖 Email Newsletters

Not everyone uses social media. So how do you engage with them? A good option is email.

Email Plugin

An email newsletter offers you the opportunity to share relevant information with your clients while building trust and your reputation. How can you do this?

NOTES

_____

_____

_____

_____

Create an email newsletter template that includes:
- Local news or current events
- News about new menu items
- Events at restaurant
- Focus on an employee, customer, new program

Farming to email addresses that did not opt-in is considered SPAM – bottom line.

The first step in your email strategy is to offer an email list sign up on your website. Before you install any plugins, you need to select an email provider for your list. Some of the online reservation systems may offer an email list component.

Current Database/CRM

Find out if your reservation system offers an email marketing program. If it does, then find out if they offer a WordPress plugin to incorporate it into your website, and if not, do they offer HTML widget code that you can place in a widget on your website?

NOTE: Keep in mind that this database should include

NOTES

_____

_____

_____

_____

people that have opted in to email. If not, then email-ing them may be considered SPAM, and you shouldn't do it.

If you do not have a current list, then decide where you want to create your list. Do you want to use a CRM or do you want to use an email program? If you select an email program, can you capture leads into your CRM and email list at the same time? You might want to use a CRM system to keep track of frequent customers, their birthdays/anniversaries, etc.

I use Gravity Forms with MailChimp. So in addition to the email signup on my website, my contact forms also feed my leads to MailChimp.

Here are some CRM and guest management programs that you may be able to feed leads to from your WordPress website or email program:

- [ ] **BUZZTABLE**
  http://www.buzztable.com/
- [ ] **FIVE STARS**
  http://www.fivestars.com/
- [ ] **POSIQ**
  http://posiq.net/restaurant-crm
- [ ] **VENGA**
  http://www.getvenga.com/

If you are not using a CRM, and even if you are, you can still choose to use an email marketing platform like MailChimp for your email marketing campaign. Some of these companies include:

NOTES

_____

_____

_____

_____

☐ **AWEBER**
http://www.aweber.com/
☐ **CONSTANT CONTACT**
http://www.constantcontact.com/
☐ **MAILCHIMP**
http://mailchimp.com/
☐ **MADMIMI**
https://madmimi.com/

Email Plugin

Once you select the email platform you want to use, you can usually create HTML code within your account that can be used to create a sign-up widget on your website. You can also search the plugin repository to see if there is a plugin for your email provider.

Once you create your sign-up widget, test it with your own email to ensure that it works. This will also put you on your email list so that you receive a copy of every newsletter you send out.

Creating a List

Think about your email strategy before you create your lists. The signup on your website may be able to offer options to your visitors. Think about it. A buyer is probably looking for different information than a seller. So offering an option, or the ability to subscibe to both might be a good idea. If you are creating your first list, this is a great time to map out your strategy. If you already have a list, you will need to determine how you will use it. For example, you can have a FREQUENT DINER call out and CA-

NOTES

_____

_____

_____

_____

TERING call out in your newsletter.

What can you manage? This is a major decision factor in setting up your email newsletter program. If you offer a quarterly newsletter, you will need to create one. You are building a website and online presence, and not everything needs need to be built at the same time. An email signup can be added later.

Email Newsletters

Depending on the email platform you selected, they will probably provide email templates that you can customize with your logo and colors. Determine what you want to include in your newsletter, and design your template around that. The purpose of your email newsletter is to:
- Establish yourself as an expert
- Increase your referrals
- Create calls to action

Make a list of the content you would like in your newsletter and then use that list to create a layout. The layout will help you decide on a layout.

Your content list might include:
- Special food items at your restaurant
- Event ideas for catering and/or restaurant events
- Frequent diner program
- Events at your restaurant
- Focus on an employee

Remember that your email newsletter is part of your branding.

NOTES

_____

_____

_____

_____

Look for a template that is consistent with your website and other marketing materials.

_____

_____

_____

_____

Content Strategy

Learn how to increae the engagement of your WordPress
website

- Content Strategy
- Editorial Calendar

Content Strategy

If you haven't gotten the message yet, let me tell you that blogging is very important to your website strategy. And one of the biggest benefits of using WordPress for your website is the blogging platform. It's ease of use along with the features it offers, will help you with your content strategy.

The first step in your content strategy is coming up with a list of topics you are already comfortable blogging about that would

NOTES

_____

_____

_____

_____

add value to your visitors. Another factor is the ability to schedule posts. So writing a blog post every day is not necessary. You can allocate a day and time for blogging and then schedule when you want them to post.

You should review your keywords and your niche focus to help you with your list of topics.

Remember that a job worth doing is worth doing well. Creating content that interests your readers and provides value will help to increase your value to them and build your reputation. They will want to come back to learn more.

5 Steps to Creating a Good Content Strategy

1. **Create titles that evoke interest.** You talk to buyers and sellers every day. Think about the questions they ask you, because there are probably great topics for blog posts in those conversations.

2. **Include SEO keywords**. Try to determine how to include your keywords in blog post titles and content. This can sometimes become the focus more than the content, and when it does you need to step away. The article should be about valuable content for your readers, not about SEO.

3. **Include photos**. You've heard it before, a picture is worth a thousand words. Remember that this photo will appear when you autopublish your post on Facebook, Google+ and any other platform you share it with. So it needs to be relevant to the headline and content of your post.

NOTES

_____

_____

_____

_____

HEADLINE IDEAS

SHARE INSIDER KNOWLEDGE

Have you heard about ... ? | The secret of ...

PROBLEM SOLVING

Here's a Quick Way to... | Get Rid of (problem) ...

CURIOSITY

What Everyone Should Know About (topic) ...

TIPS

Top 10 Tips to ...

MISTAKES

5 Mistakes to Avoid When (issue)...

NOTES

_____

_____

_____

_____

4. **Short and sweet**. No, not your blog post, your sentences. Keep the sentences short to make it easier for your readers. Use lists and/or bullets, photos and white space to make it easier for your readers. A blog post should convey a message while being easy and quick to read. You don't want to lose their interest.

5. **Call to action**. This should not be a sales pitch. In fact, it doesn't have to be a visual call to action. It can be as simple as ending your post with a question. Ask your readers for their input, their ideas, ways they solved a problem. Engage them by asking a question in the hopes of getting a response. If you do get a response, make sure you join the conversation even if it's just to say thank you. This acknowledges the reader's effort in responding.

Types of Blog Posts

There are several different types of blog posts, but if you break them down into categories, you can use them to help develop your editorial calendar. And then as you grow your following, you can determine which type of blog post creates the most engagement.

How-To Post

This is a great post type for restaurants to share menu and event planning ideas, and usually is an easy read. It is similar to the chapter on SEO where you were given the steps on how to set up a plugin along with screen captures.

- How to Plan an Event
- How to Create a Menu for Foodies
- How to Pair Food & Wine

NOTES

_____

_____

_____

_____

LIST POST

List posts are just that - lists. And they are successful in creating interest. They offer a promise of something - 7 ways to, 5 reasons, 10 tips. And that is impactful. The number doesn't matter, it's just that it's a list. If the list is long, you can break it down to more than one blog post. 10 ways to .... 5 more ways to.

- 5 Great Party Appetizers
- 10 Ways to Prepare Salad
- 6 Secrets of a Successful Pizza Party

CURATED POST

Part of becoming known as an expert is your ability to share information learned from other people. A curated post would be an example of this. Your restaurant may have just had someone come talk to you about new wine selections. Create a curated blog post on what you learned, sharing links to the wine distributor website and highlight some of their other wines that are available. You don't need to write a lot of content, because you can curate the content that already exists. A curated blog post can include video, infographics, etc. that help you share relevant information.

- New Wine Blends
- Business Growth as " " Announces New Location
- Food & Wine Festival Downtown
- Restaurant Week Participants

Editorial Calendar

Once you understand how to create a content strategy and the different types of blog posts, you should be able to come up with a list of topics that you are ready to share with your readers. I

NOTES

_____

_____

_____

_____

have found that an editorial calendar is a great way to organize my topics and help me execute my strategy. The great thing is that there are editorial calendar plugins that you can add to your website to help you keep everything in one place. If you have posts scheduled, and something comes up that you really want to blog about right away, you can just drag your scheduled post to another day and add your new post.

Editorial Calendar Plugins

- ☐ CoSchedule
- ☐ Edit Flow
- ☐ Editorial Calendar

Scheduling a post:

Here is how you schedule a post in WordPress:

1. Click Edit.
2. Change the date.
3. Change the time.
4. Click OK.
5. Save by clicking on Schedule.

NOTES

_____

_____

_____

_____

Using Photography

Learn how to use photography in your WordPress website
- 📖 What You Need To Know
- 📖 Finding Photography
- 📖 Editing Photographs

As a restauranteur, you know the importance of photography on your website.. Understanding some basic rules and how to use them will be very beneficial.

What You Need To Know

Creating original content for your website and blog can be a challenge, but it's nothing compared to the photographs. It can be a challenge to find the right photographs to use on your home page or website pages. And then finding just the right image to

NOTES

_____

_____

_____

_____

use for a blog post can take longer than writing it. But that photo is not yours. You can't use it, unless you have permission. If you do have permission, get it in writing.

The easier thing would be to take your own photos. If you take the photos, then there is no question about your right to use them.

Website photography tips:

- Do not use photos from other websites without permission, *in writing*
- Use your own photos
- Do not use photographs of food items that are not yours
- If you are using professional photographs, make sure you keep copies of your rights to the photos and watermark them
- If you use a photo service, keep copies of your agreements

You can also watermark *your* photos to claim them as yours and monitor whether anyone is using them. There are WordPress plugins that you can use to watermark your images.

- ☐ Easy Watermark
- ☐ Image Watermark
- ☐ Watermark WP Image Protect

Services that can help you with protecting your photographs:

- ☐ **DIGIMARC**
  https://dfi.digimarc.com/

NOTES

_____

_____

_____

_____

☐ **PicMarkr**
http://picmarkr.com/

☐ **Image Raider**
http://www.imageraider.com/

And just in case you thought Pinterest was a good source of images, think again. You cannot download images and upload them to your website to use as your own. And pinning images to Pinterest is a completely different topic than using Pinterest images on your website.

Finding Photography

There are several websites that offer subscriptions for royalty free photographs to use on your website. You can also hire a professional to take photos for you, or you can take them yourself. *If you hire a professional, confirm that you will have copyright of the images.* And then there are websites that offer free photography for you to use, but you need to check the Creative Commons and copyright laws.

Here is a list of websites that offer free images for you to use on your website:

☐ **Compfight**
http://compfight.com/

☐ **Death To The Stock Photo**
http://join.deathtothestockphoto.com/

Notes

_____

_____

_____

_____

☐ **EVERYSTOCKPHOTO**
http://www.everystockphoto.com

☐ **FLICKR**
http://www.flickr.com/creativecommons/

☐ **FOTOR**
http://foter.com/
WordPress Plugin: Free Stock Photos Foter

☐ **GOOGLE IMAGE SEARCH**
http://www.google.com/advanced_image_search

☐ **PIXABY**
http://pixabay.com/

☐ **MORGUE FILE**
http://www.morguefile.com/

☐ **STOCKVAULT**
http://www.stockvault.net/

☐ **UNSPLASH**
http://unsplash.com/

☐ **WIKIMEDIA COMMONS**
http://commons.wikimedia.org/wiki/Main_Page

NOTES

_____

_____

_____

_____

Editing Photos

If you are using photos from your smartphone on your website, you should edit them before you upload them to your site. These files are usually very large and higher quality than you need on your website. In addition to editing them before you upload them to your site, you can edit them even more from within the WordPress media uploader.

Once you upload an image, you can click edit to modify it.

You can crop the image by clicking on the crop tool and selecting a starting point within the image.

A dotted frame will appear over the image for you to select the area you want to keep.

Save.

You will see that the screen will brighten the area you want to keep and shade out the part you don't want. The dimensions of the image you want will also appear on the right side of the screen.

NOTES

_____

_____

_____

_____

You can also scale the image as you can see above. You only need to input one number, because the other number will be proportional to the first, and it doesn't matter which number you select, but you can only scale down - not up.

Click update to save your changes.

Notes

_____

_____

_____

_____

If you decide to edit the image once you place the image on your site, you can do that as well. Click on the image in the editor.

Click on the landscape icon that appears to edit the image from within a post. You will have editing options and Advanced Settings. In the first screen, you can change the alignment of the image, modify the link, add a Title, Caption or any of the boxes listed.

Save your changes.

Click on Advanced Settings to make more modifications including adding borders, and other CSS changes you may want on this particular image as you can see on page 170.

NOTES

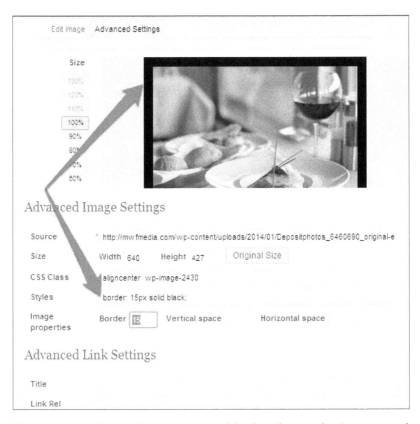

You can see above that you can add a border to the image, and stylize it as you want with CSS. Some style ideas can be found here at http://www.w3schools.com/css/css_border.asp.

NOTES

Reputation Management

&& CHAPTER &&
18

Learn how to:
- 📖 Monitor your online reputation
- 📖 Manage your online reputation

Monitoring your online reputation

We are creatures of habit and tend to not worry about anything until it becomes a problem. Social media enables anyone to say anything. Filters aren't always required, and barometers of truth and integrity don't necessarily prevail. That said, you need to monitor what is being said. Review sites enable customers to comment on your services without your consent, and don't always allow you to respond.

NOTES

_____

_____

_____

_____

And as you have learned in other chapters of this book, frequent updates are received quickly by search engines. It is quite possible for your restaurant to appear in a negative review before your website appears on a search result.

You can use several tools to monitor your online reputation. These tools offer an opportunity to automate the monitoring process so that you receive alerts whenever anything is said and can be proactive in remedying the situation. Some of these tools are relevant to your personal online reputation while others include your restaurant reputation.

DEFINITION
According to Wikipedia: "Reputation management is the practice of understanding or influencing an individual or business' reputation."

NOTES

_____

_____

_____

_____

☐ **ME ON THE WEB**
http://google.com/
Google offers a tool within the Google Dashboard, that
can help you understand and manage what people seè
when they search for you on Google. You can choose to
be notified when your personal data appears on the web
and also get tips on removing content from Google's
search results.

☐ **NAYMZ**
http://www.naymz.com/
Naymz offers online brand management and calculates
your influence across LinkedIn, Facebook, and Twitter. It
also offers rewards for members with strong reputation
scores.

☐ **REPUTATION**
http://www.reputation.com/
This website offers personal and business monitoring
through a dashboard that quickly offers you insight into
your online reputation. While their personal monitoring
is free, their other services are not.

Manage your online reputation

Much like your personal reputation, you want to ensure the integrity of your business. You want to ensure that you "own" your
online identify. Think about your business name.

Scenario: You came up with a great name for your busi-

NOTES

_____

_____

_____

_____

ness and the .com domain wasn't available. So you went ahead and purchased the .net domain. After your business launched, the .com website went viral and monopolized the internet - only problem is they are an unethical company located in the same state. And now you are suffering the negative impact. Oops!

Scenario: All of your social media efforts have paid off and you are now regarded as a subject matter expert on Twitter. The problem is your Twitter profile name isn't your real name and you never updated your profile to reflect the "real" you. And your other social media platform names are different as well.

As you begin the adventure, consider the journey. What do you want to achieve? Who do you want to be? We mentioned earlier in the book to be consistent in your online presence, and suggested you use Namechk to find out if your "name" is available on the platforms you want. You should also register your name on platforms you aren't currently active on, but may be in the future. As you continue to build your online brand reputation, you want to protect it.

Follow the rules

The ability to place an online review has become a hot button. Companies can be forced to pay a fine for placing fake positive reviews on the internet. And as they are trying to remedy a prob-

NOTES

_____

_____

_____

_____

lem, you can sometimes create an even bigger one. Companies that have abused review policies can also easily be found on the internet. Their website content over the last few years would probably have moved the negative reviews far down in the search results, but instead they are still suffering the consequences of their unethical actions.

Find out how to remedy a bad review and do it ethically. Google offers tips on handling negative information. Many review sites offer you the ability to respond to negative reviews. While it's never nice to hear something negative, it does present an opportunity to improve your services and learn something. A positive response from you to the negative review offers others an opportunity to see that you are open to improving your business and learning from your community to offer them something better.

Remedy the situation

You can score points in the way you manage negative feedback. And just like developing your online presence, it begins with listening.

1. **LISTEN** - listen to what the reviewer has said. Hear what they are saying.

2. **COMPASSION** - understand what they are saying and why they are saying it. Also consider the situation and develop a plan to action for resolution and reward.

3. **RESOLVE** - reach out to the reviewer to discuss the review and

NOTES

_____

_____

_____

_____

determine how to resolve the problem. Don't disregard their review, it is a learning experience both in what happened, but also how to handle this type of situation.

4. **REWARD** - the reviewer offered you an opportunity to improve your business and services. Offer them a reward. This does not have to be monetary, it can be a "thank you", and you hopefully considered and planned for this in step 2.

5. **REVIEW** - how did the situation occur? What could have been done differently to prevent this? Institute any new policies and procedures necessary to prevent it from happening again.

6. **SHARE** - be transparent. Share what you can about the situation through your blog and on social media to present a positive spin on the situation. This should also be done in a timely manner to present your business in a positive manner.

Restaurant
DETAIL

Create an online reputation management policy for your business and keep it updated, Also ensure that you have a Privacy Policy for your website. Protect account access and passwords, including changing passwords regularly and immediately deleting passwords and account access for employees that leave or are dismissed.

NOTES

_____

_____

_____

_____

Restaurant DETAIL

Restaurant review sites are among the most popular review sites on the internet and available on smartphones. Many users visit these apps to help them make restaurant selections. So ensuring that you are monitoring your online reputation and responding to negative reviews as well as engaging your community can have lasting results.

As more consumers search Yelp and other online reviews sites for restaurant reviews, it becomes more important for your business to retain control of your online presence and reputation. Offering quick access to post Yelp reviews and others on your website can increase your online reviews as well as help you monitor your online reputation.

You can also add your Yelp reviews to your website with a plugin.

- ☐ Yelp It
- ☐ Yelp Reviews Ticker
- ☐ Yelp Widget Pro

It is your responsibility to monitor your reputation, online and off. But by having access to online reviews, you can quickly respond to both positive and negative reviews, enabling you to remedy a bad experience. Don't ignore poor reviews, offer a res-

NOTES

_____

_____

_____

_____

olution. Do not engage in a negative conversation online, but do respond to all reviews in a positive manner.

_____

_____

_____

_____

Website Optimization

Learn how to improve the performance of your WordPress website

- 📖 Testing your website performance
- 📖 Using a CDN
- 📖 Using W3 Total Cache

Technology

"Fast is better than slow"
- one of Google's 10 things they know to be true

NOTES

_____

_____

_____

_____

Testing Your Website Performance

One of the things that can hurt your search engine results and cause visitors to leave your website is the speed at which it loads. As you add content, images, pdf files, and plugins it begins to impact your website performance. Fortunately, there are ways to test this and react if needed.

Browser Testing

Before you check for performance, you should test your website in each browser to make sure that the results are consistent. This includes Chrome, IE8 or greater, Firefox and Safari. If your site is responsive, you should also view it on an iPad and smartphone. You can test for responsive features here:

☐ **BROWSERSTACK**
http://www.browserstack.com/responsive

☐ **RESPONSINATOR**
http://www.responsinator.com/

Website Performance

There are certain things you can do to quickly to enhance your website performance. You can implement these before or after you test your website performance.

1. **DELETE YOUR SPAM COMMENTS.**
Even though you have a SPAM prevention plugin, there are probably still SPAM comments on your site. Checking this regularly and deleting them is a good maintenance practice,

NOTES

_____

_____

_____

_____

2. **DELETE UNUSED PLUGINS**

As you are setting up your website, you might be installing plugins, testing them and then deciding you don't like how they work. But you really need to deactivate and delete them. Since they are still activated on your site, they could be impacting your website performance.

3. **DELETE POST REVISIONS**

Every time you modify and save a page or post, WordPress creates a revision. If you are editing a post, it would be a good practice to preview your changes as you are working and save when you are done. If you are more comfortable with saving as you modify, then be prepared to clean up post revisions on a fairly regular basis. To do this you will need to use a plugin.

- ☐ Revision Cleaner
- ☐ Thin Out Revisions
- ☐ WP Document Revisions

You can also use ManageWP, which is a premium plugin that offers this functionality with the click of a button.

- ☐ **CLEAN-UP DATABASE**

As you continue to modify your website, adding content and more, you are adding more "junk" to your database. So a periodic clean-up will help to improve your website performance. And since you are not the only one with this problem, there are plugins that can help you with this.

NOTES

_____

_____

_____

_____

☐ WP Clean Up
☐ WP-DB-Backup
☐ WP-Optimize

As we have stated before, any time you install or delete plugins and/or themes, be sure to back-up your site before-hand in case you need to restore your site.

4. **CACHE YOUR SITE**
   Caching your website can significantly improve your web-site performance and I will provide directions on how to use two of these later in the chapter. *Note that using one of these plugins will probably replace you needing to update your htaccess file.*
   ☐ Quick Cache
   ☐ W3 Total Cache
   ☐ WP Super Cache

Testing Your Site

So now you are ready to test your site. A website performance test should provide information on site speed as well as page sizes and response. The test should also include recommendations on ways to improve your website performance.

The website performance site I use is GTMetrix because they provide a WordPress Optimization guide and a WordPress plugin to help me monitor my website. But there are others:

- **GTmetrix**
  http://gtmetrix.com/
- **Pingdom**
  http://tools.pingdom.com/fpt/
- **WebPageTest**

NOTES

_____

_____

_____

_____

http://www.webpagetest.org/
The YSlow browser extension can also be used for page speed, which is available at http://developer.yahoo.com/yslow/. The Google Page Speed is available at the Chrome Store (https://chrome.google.com/webstore/search).

Once you run your performance test, you will know where there are opportunities to improve your site's performance. While it may seem strange, you will need to install more plugins to improve your website speed and performance.

Using a CDN

Definition

A Content Delivery Network is a system of distributed servers (network) that deliver webpages and other Web content to a user based on the geographic locations of the user, the origin of the webpage and a content delivery server.

- Webopedia

So not only does a CDN save bandwidth costs from your hosting provider, but by serving information it helps to increase the performance of your website, which has the potential to increase your search results as well as improving the user experience.

NOTES

_____

_____

_____

_____

Why should you use a CDN?

There are three basic reasons for using a CDN:

1. Speed
   this isn't always easy to see when you have one image on a blog post, but if you have a portfolio page on a CDN vs. your WordPress media file, you will be amazed at how much faster the page loads.

2. Improved User Experience
   this has a lot to do with speed. The less time a visitor has to wait for a page to load, the more likely they are to stay on your site, so your bounce rate will go down.

3. Improved SERP
   again this has to do with the speed of your site. Google has stated that faster sites tend to rank higher in search results.

What can you put on a CDN?

You can use a CDN for static content which includes images, pdf files, stylesheets (css files), javascripts, etc. A static file is a file that doesn't change. They are uploaded to your server and remain in the same state unless you download them to modify and replace them.

Other ways to minimize impact of images:
- ☐ Image Cleanup
- ☐ Resize Image After Upload
- ☐ WP Smush.it

TinyPng
https://tinypng.com/

NOTES

_____

_____

_____

_____

How to use with Photoshop images:
Use Save for Web to export your images as 24-bit transparent
PNG files and upload them to TinyPNG.

Optimization Tools
Performance Optimization: Order Styles and Javascript

You may have plugins that are contributing to a slowdown on
your website. See which plugins might be slowing down your
website.
- ☐ Debug Objects
- ☐ P3 (Plugin Performance Profiler)

MONITORING TOOLS
- ☐ Google Pagespeed Insights for WordPress

MOVE SCRIPTS
- ☐ Scripts to Footer

*The following pages offer code for your .htaccess file as well as some
plugins you can use to optimize your site. It is highly recommended
that you consult a professional before you modify your .htaccess
file, as you could "break" your site. In addition, while these are
suggestions, every site is unique and you may not need some or any
of these items.*

NOTES

_____

_____

_____

_____

Always make a copy of ANY file you are modifying before you make any changes and save it in its original state. You can copy and paste this back or upload it to override any changes you made. Take it from someone who has learned this lesson - from making this mistake.

Expires Headers

Expires headers are a way of setting long cache expirations on your static files (CSS, images, JavaScript, etc.) which can help improve performance. This is the code to add to your .htaccess file.

```
# Create expires headers
<FilesMatch "\.(ico|jpg|jpeg|png|gif|js|css|swf)$">
ExpiresActive on
ExpiresDefault "access plus 30 days"
Header unset ETag
FileETag None
</FilesMatch>
# End Create expires headers #
```

NOTES

_____

_____

_____

_____

Google Page Optimization

Before you update your .htaccess with the Google Page Speed code below, check with your host on what PHP version you are using and if you can use the Google Page Speed.

```
<IfModule pagespeed_module>
  ModPagespeed on
  ModPagespeedEnableFilters remove_comments,rewrite_javascript,rewrite_css,rewrite_images
  ModPagespeedEnableFilters elide_attributes,defer_javascript,move_css_to_head
  ModPagespeedJpegRecompressionQuality -1
</IfModule>
```

Browser Caching

Place this code at the top of your htacess file.

```
## EXPIRES CACHING ##
<IfModule mod_expires.c>
ExpiresActive On
ExpiresByType image/jpg "access 1 year"
ExpiresByType image/jpeg "access 1 year"
ExpiresByType image/gif "access 1 year"
ExpiresByType image/png "access 1 year"
ExpiresByType text/css "access 1 month"
ExpiresByType application/pdf "access 1 month"
ExpiresByType text/x-javascript "access 1 month"
ExpiresByType application/x-shockwave-flash "access 1 month"
ExpiresByType image/x-icon "access 1 year"
ExpiresDefault "access 2 days"
</IfModule>
## EXPIRES CACHING ##
```

NOTES

_____

_____

_____

_____

GZip Compression

This needs to be placed at the **bottom** of your .htaccess file.
Once you update your file, you can test your compression results
at:

- ☐ http://99webtools.com/http_compression.php
- ☐ http://checkgzipcompression.com/.
- ☐ http://www.gziptest.com/

```
# compress text, html, javascript, css, xml:
AddOutputFilterByType DEFLATE text/plain
AddOutputFilterByType DEFLATE text/html
AddOutputFilterByType DEFLATE text/xml
AddOutputFilterByType DEFLATE text/css
AddOutputFilterByType DEFLATE application/xml
AddOutputFilterByType DEFLATE application/xhtml+xml
AddOutputFilterByType DEFLATE application/rss+xml
AddOutputFilterByType DEFLATE application/javascript
AddOutputFilterByType DEFLATE application/x-javascript
AddType x-font/otf .otf
AddType x-font/ttf .ttf
AddType x-font/eot .eot
AddType x-font/woff .woff
AddType image/x-icon .ico
AddType image/png .png
```

NOTES

_____

_____

_____

_____

Remove query strings

This needs to be placed in functions.php file before the closing PHP tag <?

```
function _remove_script_version( $src ){
        $parts = explode( '?', $src );
        return $parts[0];
}
add_filter( 'script_loader_src', '_remove_script_version', 15, 1 );
add_filter( 'style_loader_src', '_remove_script_version', 15, 1 );
```

Plugins
- [ ] Query Monitor
- [ ] Remove Query Strings From Static Resources

Minify Javascript and CSS

Plugins
- [ ] Better WordPress Minify
- [ ] Simple Minify
- [ ] WP Minify

That's it for modifying the .htaccess file. If you are going to use one of the cache plugins, you may not need some or any of these modifications, because the plugin will include them.

NOTES

_____

_____

_____

_____

Using W3 Total Cache

While W3 Total Cache is a popular cache plugin for WordPress websites, there are certain themes and other plugins that it may conflict with. Depending on the conflict, you may not be able to use this plugin, and may need to find another.

You should have access to your .htaccess file, FTP and host provider in case you need to deactivate and delete the plugin.

Once you have W3 Total Cache installed, you will see a new Performance section on your WordPress dashboard. This plugin is very robust, and we will only be reviewing certain settings that you should consider configuring. You can always go back and modify others later if you like.

Select Performance on your dashboard to see the different options within the plugin.

NOTES

_____

_____

_____

_____

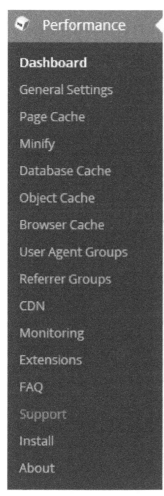

Performance

Dashboard

General Settings

Page Cache

Minify

Database Cache

Object Cache

Browser Cache

User Agent Groups

Referrer Groups

CDN

Monitoring

Extensions

FAQ

Support

Install

About

The following items should be checked, and any others should be left at their default settings. If something is checked, and not included in the information below, leave it alone.

GENERAL SETTINGS
Click on General Settings and we will begin to toggle some of the options.

Page Cache:
Enable Page Cache
Page Cache method - Disk enhanced.

Minify:
Enable minify
Minify mode set to auto
Minify cache method - Disk
HTLM minifier - Default
JS Minifier - JSMin (default)
CSS minifier - Default

Database Cache:
Enable database cache
Database Cache Method - Disk

Object Cache:
Enable object cache
Object cache method - Disk

NOTES

_____

_____

_____

_____

Page Cache

Enable page caching to decrease the response time of the site.

Page cache: ✓ Enable
Caching pages will reduce the response time of your site and increase the scale of your web

Page cache method: Disk: Enhanced ▼

Save all settings Empty cache

Minify

Reduce load time by decreasing the size and number of CSS and JS files. Automatically remove unnecessary data from CSS

Minify: ✓ Enable
Minification can decrease file size of HTML, CSS, JS and feeds respectively by ~10% on avera

Minify mode: ● Auto ○ Manual
Select manual mode to use fields on the minify settings tab to specify files to be minified, ot

Minify cache method: Disk ▼

HTML minifier: Default ▼

JS minifier: JSMin (default) ▼

CSS minifier: Default ▼

Save all settings Empty cache

Browser Cache:

Enable browser cache

Notes

_____

_____

_____

_____

Database Cache

Enable database caching to reduce post, page and feed creation time.

Database Cache:	☑ Enable
	Caching database objects decreases the response time of your site. Best used if object caching is not p

Database Cache Method:	Disk ▼

Save all settings   Empty cache

Object Cache

Enable object caching to further reduce execution time for common operations.

Object Cache:	☑ Enable
	Object caching greatly increases performance for highly dynamic sites that use the Object Cache API

Object Cache Method:	Disk ▼

Save all settings   Empty cache

Browser Cache

Reduce server load and decrease response time by using the cache available in site visitor's web browser.

Browser Cache:	☑ Enable
	Enable HTTP compression and add headers to reduce server load and decrease file load time.

NOTES

_____

_____

_____

_____

PAGE CACHE

General:
Enable Cache front page
Enable Cache feeds: site, categories, tags, comments
Enable Cache requests only for [mydomain.com] site address
(where [mydomain.com] is your domain)
Enable Don't cache pages for logged in users

General

✓ Cache front page
For many blogs this is your most visited page, it is recommended that you cache it.

✓ Cache feeds: site, categories, tags, comments
Even if using a feed proxy service (like FeedBurner), enabling this option is still recommended.

☐ Cache SSL (https) requests
Cache SSL requests (uniquely) for improved performance.

Cache URIs with query string variables
Search result (and similar) pages will be cached if enabled.

☐ Cache 404 (not found) pages
Reduce server load by caching 404 pages. If the disk enhanced method of disk caching is used, 404 page

✓ Cache requests only for bonniedupree.com site address
Cache only requests with the same URL as the site's site address.

✓ Don't cache pages for logged in users
Unauthenticated users may view a cached version of the last authenticated user's view of a given page.

☐ Don't cache pages for following user roles
Select user roles that should not receive cached pages:

Administrator   Editor   Author   Contributor   Subscriber

[Save all settings]

NOTES

_____

_____

_____

_____

MINIFY

General:
Enable Rewrite URL structure
Enable Minify error notification to Admin Notification

HTML & XML:
Enable HTML minify settings
Check Inline CSS minification
Check line break removal.

NOTES

_____

_____

_____

_____

General:
Enable Set Last-Modified header
Enable Set expires header
Enable Set cache control header
Enable Set Entity tag (eTag)
Enable Set W3 Total Cache header
Enable HTTP (gzip) compression

Once you have completed all of these settings, go back to the Performance menu.
- Click on the button at the top to EMPTY ALL CACHES.
- That's it.

NOTES

_____

_____

_____

_____

Using WP Super Cache

Once you install WP Super Cache, you will need to activate the plugin in order to set it up.

> **Plugins**
>
> WP Super Cache is disabled. Please go to the plugin admin page to enable caching.

Turn Caching on and update.

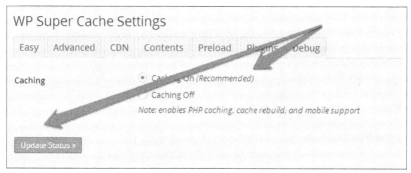

Now you can move on to the Advanced tab.

NOTES

_____

_____

_____

_____

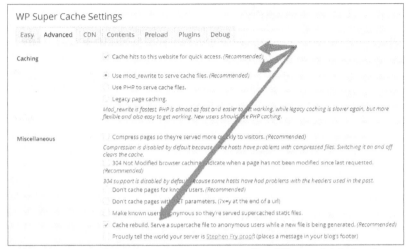

Select **Use mod_rewrite to serve cache files**, which is also recommended.

And under Miscellaneous, select Cache rebuild (*also recommended*).

Update status. Scroll further and update mod_rewrite which should turn the code box green, signifying you successfully updated the code.

NOTES

Technology

Make note of the uninstall directions on this page if you decide to use a different cache plugin or not use this one.

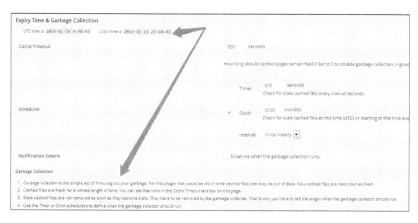

Next you should set up a time for expiry and garbage collection. Read the content at the bottom of the page for an explanation and set a time for expiry.

NOTES

_____

_____

_____

_____

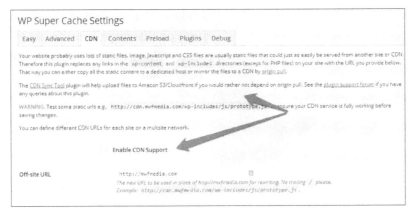

If you are using a CDN, follow the directions to enable WP Super Cache with a CDN.

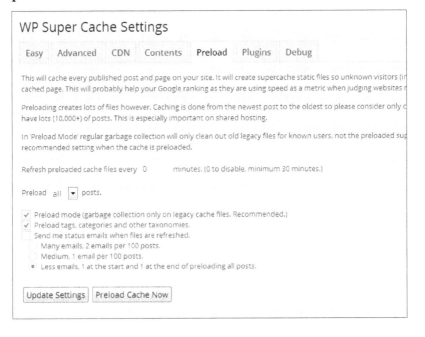

NOTES

_____

_____

_____

_____

Preload would also help your website optimization, but your settings would depend on your particular website. Follow the directions on the page.

That's it for some basic website optimization. These are suggestions, but many of them can severely impact your website, so tread lightly. Here are just a few things to keep in mind:

1. Always back-up your website before you make any modifications.
2. Ensure you have access to your FTP and host in case you need to delete or restore.
3. Make copies of all of your files in a text editor before you modify them. Save them in their original format in case you need to restore them later.

Some host companies offer installation directions for these plugins using their host. Check if yours does - for example:
- *https://support.hostgator.com/articles/specialized-help/technical/wordpress/wp-super-cache-plugin*
- *http://www.bluehost.com/blog/educational/websites/the-ultimate-guide-to-supercharging-your-wordpress-blog-2097/*

NOTES

_____

_____

_____

_____

NOTES

Backup & Security

Learn the importance of backing up your website and integrating some security measures.

- 📖 Backing up your website
- 📖 Integrating security measures

Backing up your website

Many businesses are so involved in setting up their websites, analytics and SEO, that they forget about backing up their site. Using a backup plugin or service is an important step in your website maintenance plan. There are usually backup plans available through your hosting provider, as well as plugins on the repository and premium plugins. Understanding your particular situation will help you determine the best solution. I cannot impress on you enough how important a backup strategy is. I have heard

NOTES

_____

_____

_____

_____

all too often about a website crashing, and the owner in a panic because they never backed it up.

As you develop your backup plan, you need to understand the difference between a site backup and a database backup. Your WordPress installation consists of at least:

- WordPress Core Installation
- WordPress Plugins
- WordPress Themes
- Images and Files
- JavaScript and PHP scripts, and other code files
- Additional Files and Static Web Pages

Your WordPress database includes your posts and data, but not all of the above. Most of these items are what creates your website from its design to the information on your site. So you need to backup both.

How often should you backup?
If you do not update your website on a regular basis, then creating daily, or even weekly backups is probably not necessary. On the other hand, if you blog frequently and add new content on a regular basis, then you should consider a daily backup.

How many backups should you keep?
WordPress recommends three backups in case a file becomes corrupt or damaged. Depending on your backup storage system, and the size of your backup, this can become costly, but should be thought of in the same way as an insurance policy.

NOTES

_____

_____

_____

_____

Can I automate my backup?

This will depend on the plugin you select, but it is recommended that you automate your backups. One day of forgetting can be the day your site crashes after you spent hours updating and adding content. Your backup plan should include steps for automating and restoring your backup, as well as where your backups will be stored.

Backup Plugins

- ☐ DBC Backup 2
- ☐ UpdraftPlus - WordPress Backup and Restoration
- ☐ XCloner - Backup and Restore
- ☐ WP-DBManager

Premium Backup Plugins

- ☐ **Backupbuddy**
  http://ithemes.com/purchase/backupbuddy/

- ☐ **CodeGuard**
  http://blog.codeguard.com/codeguard-wordpress-back-up-walkthrough/

- ☐ **ManageWP**
  https://managewp.com/

- ☐ **VaultPress**
  http://vaultpress.com/

- ☐ **WPMU Dev Snapshot**
  https://premium.wpmudev.org/project/snapshot/

NOTES

_____

_____

_____

_____

Integrating Security Measures

Even though you are taking steps to backup your website, you also need to implement security measures to protect against hackers. These measures range from spam protection to login restriction to strong security programs.

Security Measures

1. Make sure that your host is backing up your website. Even though you will be backing up your website, having a website backup on your host server can expedite a restore.

2. Ensure that you have regularly scheduled backups of your website and test a restore periodically to make sure that these are working. You should also be sure that you have the ability to manually backup your site before you update any plugins or themes, and of course a WordPress update.

3. Use strong passwords. We have already discussed this earlier in this book, but it bears repeating. Do not repeat your passwords. Create unique, strong passwords for your website.

4. Do not use Admin. Again, we discussed this earlier in this book. If you use Admin for your user name, you have given the hackers 50% of the information they need to access your site.

5. Delete or rename install.php, upgrade.php and readme.html files from your WordPress installation. Once you have installed WordPress, these files are no longer needed.

NOTES

_____

_____

_____

_____

6. Always deactivate unused plugins and then delete them. Don't keep unused plugins on your site. If you don't need them, delete them.

7. Activate Akismet or use another SPAM protection plugin. Delete SPAM regularly.

8. Clean up your site of unused plugins, post revisions, spam and database overhead. This can also help to increase your website performance.

Your .htaccess File

You can place several security measures in your .htaccess file. Be aware that modifying your .htaccess file can also "break" your site if not done properly. So be sure to backup your site before you modify the file and have access to your FTP or host provider to restore the file in case something goes wrong. There are also plugins on page 218 that offer .htacess protection.

PROTECT WP-CONFIG FILE

```
# protect wpconfig.php
<files wp-config.php>
order allow,deny
deny from all
</files>
```

NOTES

_____

_____

_____

_____

DISABLE HOTLINKING

Hotlinking is when other website owners use your images and/or videos by hotlinking to them via the link on your website which can put a strain on your website server and bandwidth.

Definition

```
RewriteEngine on
RewriteCond %{HTTP_REFERER} !^$
RewriteCond %{HTTP_REFERER} !^http(s)?://
(www\.)?YourDomain [NC]
RewriteRule \.(jpg|jpeg|png|gif)$ - [NC,F,L]
```

DISABLE BROWSING

Similar to hotlinking, eliminate the ability for visitors to browse your WordPress directories.

```
# disable directory browsing
Options All -Indexes
```

NOTES

_____

_____

_____

_____

PROTECT THE .HTACCESS FILE
```
<files ~ "^.*\.([Hh][Tt][Aa])">
order allow,deny
deny from all
satisfy all
</files>
Disable Directory Browsing
```

Site Malware

Malware can not only wreak havoc on your website, but can also cause it to be shutdown (*at least temporarily*) by Google and others, as well as your host provider. While you may be using malware protection on your computer, this usually only protects your computer and not your WordPress website installation.

Malware Plugins
☐ Anti-Malware (Get Off Malicious Scripts)
☐ Sucuri Security - SiteCheck Malware Scanner
☐ Wordfence Security

Website Security

There are several plugins that offer a variety of security protections from SQL Injection hackings, account login and .htaccess protection so that you do not have to modify your .htaccess file.
☐ All In One WP Security & Firewall
☐ Better WP Security
☐ BulletProof Security
☐ Wordfence Security

NOTES

_____

_____

_____

_____

Premium Security Monitoring

There are several premium security plugins for WordPress that offer additional features. Here are some of them:

- **Acunetix WordPress Security Plugin**
  https://www.acunetix.com/websitesecurity/wordpress-security-plugin/

- **Sucuri**
  http://sucuri.net/wordpress-security-monitoring

- **Wordfence**
  http://www.wordfence.com/

FIREWALL PROTECTION

Since you are probably using your phone over wifi networks, you should have firewall protection on your smartphone and tablets similar to the one you use on your laptop and other computers.

- **Comodo**
  http://personalfirewall.comodo.com/

- **Eset Smart Security**
  http://www.eset.com/us/

NOTES

_____

_____

_____

_____

INDEX

The following pages offer:
- Index
- Websites
- WordPress Plugins

Index

A

B

C

Notes

_____

_____

_____

_____

D

E

F

NOTES

G

H

I

NOTES

_____

_____

_____

_____

J

K

L

M

NOTES

_____

_____

_____

_____

N

O

P

NOTES

_____

_____

_____

_____

R

S

T

NOTES

U

N

W

X

Y

NOTES

_____

_____

_____

_____

WEBSITES

NOTES

_____

_____

_____

_____

E

F

G

H

Notes

_____

_____

_____

_____

I

J

K

L

M

N

O

NOTES

_____

_____

_____

_____

P

Q

R

S

Notes

_____

_____

_____

_____

NOTES

WORDPRESS PLUGINS

NOTES

_____

_____

_____

_____

Database Cleanup

Editorial Calendar

Events

Facebook

Flickr Plugins

Google Plus

NOTES

_____

_____

_____

_____

Image Reduction

LinkedIn

Malware Plugins

Minify Javascript and CSS

Monitoring Tools

Move scripts

Online Order Plugins

Operations

NOTES

_____

_____

_____

_____

Page Builder

Pinterest

Plugin Performance

Policy

Post Revisions

Recipes

Remove Query Strings

NOTES

_____

_____

_____

_____

Restaurant Menu Plugins

Reservations

SEO

Social Share

SPAM Preventer

Twitter

NOTES

_____

_____

_____

_____

Watermark

Website Security

NOTES

_____

_____

_____

_____

NOTES

_____

_____

_____

_____